TUNNELLING HISTORY AND MY OWN INVOLVEMENT

TUNNELLING HISTORY AND MY OWN INVOLVEMENT

by

Sir Harold Harding
D.Sc. B.Sc. F.Eng. F.I.C.E.
Past President of the Institution of Civil Engineers

Part I: TUNNELLING HISTORY

Part II: TUNNELLING EXPERIENCE

GOLDER ASSOCIATES, TORONTO
1981

170 Attwell Drive, 6th Floor
Rexdale (Toronto), Ontario
M9W 5Z5

ISBN 0-9690971-0-7

Printed in Canada by
UNIVERSITY OF TORONTO PRESS

FOREWORD

Sir Harold Harding's experience in tunnelling, particularly *soft-ground* tunnelling, is not unique, but it is rare. What is unique is the combination of his experience, his historical memory, his sense of humour and his simplicity of expression and complete lack of pomposity.

Having worked with him for a number of years (although his junior by some ten years, he never gave me the impression that I was working under him) and knowing that he was capable of recording his experience in an instructive, interesting and amusing manner, I suggested to my colleagues that we should ask him to allow us to publish his record of his engineering experience. Wise beyond their years they agreed. This book is the result!

The purpose of a *Foreword* is to introduce the author to the reader.

I first heard of Harding while at the Building Research Station in 1939 when Dr Skempton (Skem as he then was) had been to an Imperial College dinner. "There was a fellow there called Harding — he made a speech — he told some *wizard* stories!" (Wizard dates it!)

Harding's stories are not, and never were, re-hashes of stories he had heard elsewhere and remembered; they were new creations by him — a twist of words or of a situation, which, with his impish humour became uproariously comical. As an example, Skem said that at the Engineering Dinner, Harding had related that his Welsh uncle who was a Rector, generated nineteen children with the aid of his two wives with whom he was connected "in series and not in parallel!".

Harold John Boyer Harding was born in January 1900. Throughout his life this has been a great help to him in knowing at any time how old he was! In fact his arithmetic is impeccable. But he has a dislike for the Higher Mathematics. On receiving a paper full of equations he would pass it on to me and say "Have a look at it for me Golder, it's got π in it!"

He served as an officer cadet in the 1914–18 war, from January 1918 to January 1919 and held a commission for one day. Consequently, he was at college (City and Guilds) with a lot of ex-service types, many of whom later became notable if not notorious! This is where he lost his hair! This caused him later in life to bear a strong resemblance to the Archbishop of Canterbury (Dr Fisher). His appearance has not changed since, but the Archbishop has.

There will be many readers who never had the privilege of hearing Sir Harold give a lecture, or join in a discussion at the Institution of Civil Engineers (The Civils).

When his name was called, a stir went through the audience, faces lit up, aching bottoms were forgotten, and we settled down for a treat. He was not an orator. He spoke pleasantly, in a clear and audible voice, but in the same manner in which he talked to a small group of friends in a room. Although he sometimes had notes, he seldom referred to them, possibly because his handwriting is illegible! But there was no hesitation in his delivery.

His subject matter was sprinkled with delicious original phrases which became knows as *Hardingisms*. Although it is hard to remember any actual jokes which he recounted he was interrupted often by roars of laughter from his audience. He never overstayed his welcome on the podium, but he never dried up entirely.

Some years ago he made a notable contribution to the science of Geotechnology in the form of *Harding's Law*. In response to a request for a three-ton hammer to be sent out for pile-driving in Mombasa harbour, Harding sent two. Replying to the irate protest about cost he wrote: "If you are working over water and you only have one hammer, sooner or later you will drop it in the drink. If you have two hammers you will never drop either of them in the drink. This is a well known scientific fact!" — Harding's Law.

Sir Harold has a *historical memory*. I have often heard him recall from memory long statements of the facts which now form sections of his chapters on *"Tunnelling History"*.

His working notes were always meticulously kept. I never fathomed how he was able to produce his *Lefax* pocketbook and turn to a page of notes relating to the subject under discussion no matter what it was.

He never gave me a difficult problem! Now I realise that when he discussed a problem with me and asked my advice he had already solved the problem in his own mind and wanted me to "put the soil into it Golder" — as he once said in an unguarded moment!

On one occasion I was troubled by an engineer who had crossed the Atlantic, taken a degree in *Soil Mechanics* and come back to England to tell us, who had no degrees in Soils, how everything should be done! Harding said to me "send him to me Golder and I will put him in a deep hole in the ground and teach him humility!" It worked, and proved the young man's mettle. He said to Harding "Sir, when you are at the bottom of a cofferdam 70 feet deep, it looks very different from what it did on the drawing board".

Sir Harold treated his seniors, contemporaries and juniors kindly, and all as equals, but he could be harsh when needed. On one occasion he reduced one very tough young engineer to tears by a verbal thrashing. On another, a leisurely Oxford graduate jumped out of his skin and to attention when, having draped his body over Harding's desk without permission the usually mellow voice snapped out like a whiplash!

He was horrified when on one occasion I told him that the typists were terrified of him. "But Golder, I am the mildest of men" he protested. "Mild steel" yes! working well within the elastic limit at all times!

Sir Harold was President of the Institution of Civil Engineers in 1963–4, and was knighted in 1968 for his services to Civil Engineering including his work on the Channel Tunnel.

H.Q. Golder

Contents

Part II: TUNNELLING EXPERIENCE

List of Illustrations

PART ONE

PART TWO

Part I: TUNNELLING HISTORY

INTRODUCTION

There are said to be three forms of intelligence — Animal, Human and — Military and this classification applies to tunnelling.

Burrowing animals have evolved themselves into perfectly equipped tunnelling mechanisms. Their intelligence leads them only to tunnel where the ground is suitable and to adopt a circular section.

Our human anthropoid ancestors were ill-equipped. "A life of swinging through trees was aided by retention of four fingers and an opposite thumb, by long strong arms and by acute vision that co-ordinated the use of arms and hands. That Man's ancestors forsook such an existence is attributed at least to climatic changes that replaced much of the forest with grasslands. In such an environment only a ground-running animal could survive" (Larousse p. 404).

So Homo, later to be called Sapiens, had to use intelligence and develop tools to overcome his physical shortcomings in order to make shelter and to develop protective clothing for his delicate hide. As his skills increased they were inevitably used for offence and defence against his fellow-humans. So throughout History the Military has run parallel with the Civil Engineering.

Thus it seems appropriate to quote from the address by General Sir William Jackson to the Centenary Meeting of the Institution of Royal Engineers — "On our Sapper Heritage" (Inst. R.E. Jour. 1976): "I am probably addressing several *Henry Fords* to whom History is Bunk, but I think that most of us will agree that, although history cannot be used to forecast the future, it is a data-bank of our inherited experiences and its myths and legends are our inspiration".

The term Civil Engineer was adopted in the eighteenth century to distinguish Civil from Military works. But in its widest sense it can include mining for peaceful purposes. So let us separate *Tunnellers* from *Miners* in order to simplify later references.

Consider a paradox. The object of the Tunneller is to make holes for a definite purpose of designed shape and length. The spoil which he digs is a necessary nuisance and its disposal is an unavoidable expense. The whole object of the Miner is to make use of the spoil which he digs, and to recover it is the reason for his existence. The hole which he leaves is a necessary nuisance, of a shape dictated by the veins which he is exploiting. It is abandoned when possible. Tunnellers and miners have much in common over methods of excavation, but their other problems can be in contradiction to each other so that their vocabularies differ. Their work below ground shows likenesses and contrasts and includes shaft sinking, which for historical purposes can be treated as vertical tunnelling.

The Germans have two useful short words. Der *Baugrund* is the ground in which the work is carried out. Der *Grundbau* is the work in the ground. As tunnelling grew, within the limitations of the early tools both tunnellers and miners learned the extraordinary variety of the *Baugrund* and the need to vary their *Grundbau* to reach limited success by intelligent use of hands and tools, and the ability to protect themselves with supports. They learned the lesson, unknown to ignorant laymen, that, in spite of pious hymns "the Earth beneath and the Waters under the Earth" are basically malevolent to Man. This led to the thought that "the best engineers are those who treat the forces of nature with humility until they are sufficiently sure how they can outwit them and convert them unsuspectingly to the use and convenience of man" (Harding 1964).

The history of tunnelling resembles the path of a firework rocket. A long thin trail gradually accelerates the rocket until there is a sudden burst which throws out separate trails which themselves explode in a countless shower. The first burst was caused by the Industrial Revolution, steam power, and railways. Subsequent bursts were fired by several fuses. Electric power and light made underground railways possible and as Power Stations multiplied, cooling water and cable tunnels followed. Hydro-electric power formed a varied explosion of its own. The internal combustion engine (gas, petrol and diesel) not only greatly extended the power of the engineer but led to the urge for motor transport to abandon the habit of following roads over hill and dale and to demand large road tunnels, not only under mountains but even through lesser hills if they lay in their path.

The increase in world population and the rapid growth of cities produced even more needs and purposes for tunnels, such as drains, sewers, aqueducts, water supply, cable tunnels and other uses. The need for water led to ever bigger dams needing tunnels, sometimes during construction for diversions, sometimes permanently.

Paradoxically, after aircraft appeared, the higher they soared upwards the deeper the tunneller had to dig to make shelters, underground factories, storage and command posts as Man followed his habit of turning useful inventions to his own destruction. Atomic Power Stations produced even more proliferation in cooling-water works and storage of the radio-active waste underground, as well as driving more and more of the essentials of life deeper into the ground to protect us from our own follies.

In order to disentangle the history of so many types of tunnel in rock or in soft ground one must follow main types and trends separately, although they run parallel with each other.

No report is complete unless it covers Who, What, Where, When and Why. To which can be added — How. What, When, Where and How depend upon decisions of private enterprise, Bureaucracy or Dictatorship. The credit for Who can be spread over a number of people. Why follows the many needs and purposes mentioned above.

All of this cannot be detailed in a short memoir, but anyone who is sufficiently involved to need greater detail would do well to study the 30 page Bibliography at the end of *Tunnels and Tunnelling* (Pequignot 1963). This gives references to published material under headings and sub-divisions and also in order of date which helps the historical sequence.

Chapter 1

EARLY TUNNELS TO THE INDUSTRIAL REVOLUTION

In Man's early form "the earliest fossil that can be classified as an ape-man is that of Australopithecus of South Africa. The remains date back about one million years. His brain-case was about 600 ccs in size, just a little larger than that of a gorilla. He walked upright, produced tools of bone and horn and lived in caves" (Larousse p. 404). As his brain-case increased in size he was able to improve his tools and enlarge and alter his cave dwellings. When driven to more open areas, devoid of caves, he was compelled to dig out his own shelter. When he first tried to tunnel in loose, unstable ground, his sorrowing relatives would have deduced the need to devise supports in such ground, so tunnelling in the true sense began.

The exact dating of pre-historic and bronze age workings is of less importance than the lessons and logic which derive from them. Their ages and advances vary from country to country (Derry & Trevor 1960). From the Paleolithic (20,000 BC) to the Neolithic Age (3,000 BC) tools were basically made of flint, bone or wood. In Norfolk these tools were used to sink shafts in order to mine for even better beds of flint for tool making. One can speculate on what site investigation revealed the existence of these beds. The full development of the Bronze Age (2,000 BC) and the early Iron Age (1,000 BC) led to the making of metal tools with all the benefits that this conferred. So mining for metals became wide spread. Timbers could be cut to shape to provide supports.

The Royal Tombs of Thebes in Egypt and Ur in Mesopotamia show the advanced cultures of the period. They are dated about 2,500 BC, by which time in Egypt, miners were recognised as craftsmen. From that time, until the Nazi and Russian Labour camps, tunnels and mines were worked by prisoners or slaves who were expendable, but experience proved that craftsmen were needed to oversee the unskilled labour.

Throughout tunnelling history the past and its methods keep repeating themselves. In spite of modern sophistication, primitive methods are sometimes still unavoidable.

The proliferation of underground tombs was due to the desire of a powerful ruler to immortalise himself, rather than to work *for the use and convenience of Man*. The tunnelled tombs in the Valley of the Kings at Thebes reached their zenith around 1,500 BC. In this particular limestone in a dry climate, it was possible to make long complex tunnels and chambers completely unlined which remain to this day as they were built. Wall paintings in these tombs show examples of how blocks of stone were quarried by splitting them by inserting wooden wedges which expand when wetted — an early example of the plug-and-feather. Equally surprising is the fact that about 3,000 BC, blocks of stone for pyramids were quarried in tunnels driven upwards from a cliff face for several hundred yards.

Another case where the ground was not *malevolent* is the underground Temple at Abu Simbel in Nubian Sandstone which stood for thousands of years until its monuments were re-erected above the waters of the Aswan High Dam. This brings to mind the vast underground rock temple built by General Franco in the Guadelarrama Mountains in unsupported rock in memory of the Civil War dead — and himself. Where the rock is self-supporting and dry, such vast diggings are not really great feats of tunnelling compared with many smaller and less spectacular tunnels, carried out against every conceivable handicap before the days of modern aids.

In the course of time men realised the interaction in most ground between excavation and lining, and this must still be paramount in the mind of the Engineer today.

While the Egyptian tomb industry flourished, a more primitive tomb was being used in Brittany and Great Britain consisting of Dolmens. A Dolmen is basically a flat stone resting upon two vertical stones. These would be built in line to form a covered way and then the burial mound would be filled above them, making an early form of cut-and-cover tunnel leading to the burial chamber. This resembles the timbered Box Heading, used in England up to the present day in heavy clays. In large headings these were driven using 12 inch by 6 inch head trees and verticals or side trees.

The basic principles of hard rock tunnelling and mining were fully understood in the Middle East before the Greeks and Romans became the arbiters of the Art in Western Europe.

The ability to break down a rock face by building fires against the face may have been invented by *casual observation* by cave dwellers who saw the effect of their fires on the walls of their caves. This method was pursued through the ages and even reported in Sweden in the last century.

For over two thousand years the general principles of tunnelling changed little. Mining far outstripped tunnelling and gave continuity to the science. For many years before the Roman invasion, mining for copper and tin was carried out in Cornwall and for lead in Derbyshire and Yorkshire, while the search for metals, precious or otherwise was going on all over the world in some fashion. The map of Europe is covered with hundreds of known sites of Roman mines.

One quarry for information is the monumental work by Henry S. Drinker, whose first edition came out in 1878 at a critical phase of tunnel development. It is entitled *Tunnelling, Explosive Compounds, And Rock Drills* with a long subsidiary title of twelve lines, and covers 1,500 pages. Some of his comments are very quotable.

A number of Roman works are mentioned by him and others, and found in the writings of Livy. As an example: A tunnel driven to drain Lake Albanus was 6,000 feet long, 25 feet wide and 30 feet high. It was driven in 359 BC from 50 shafts, 120 feet apart, hand driven in hard lava in one year. Some shafts were 400 feet deep and some sloping. The distance apart is much the same as on many early British railway tunnels.

Another tunnel was driven about 52 AD between Lake Fucinus and River Liris and was cleaned out by the Neapolitan Government. It was found to be about three miles long, nine feet wide and 19 feet high. It is said by Livy to have been driven from 22 openings, taking 11 years and consuming 30,000 men. They would be mostly slaves.

Setting out straight tunnels was no insuperable problem by the use of the plumb bob, the water level and the knowledge of the principles of Euclid. Probably geometrical and graphical methods were preferable to filling field books with calculations in Roman numerals.

The Roman period was prolific in tunnels for drainage and water supply, and masonry was often used for lining. Little is written of their detailed methods and of soft ground work. Through the ages timber has naturally been the first choice for temporary support. When used for permanent support its permanence is so dubious that remains are not so easy to find.

In broken rock and in mining, random propping was much used, sometimes developed into systems. The old established pit props of timber have been increasingly replaced in British coal mines by recoverable steel jacks, but in South Africa there are vast plantations of gum and fir in the Drakensberg to make timber cribs for the gold mines.

After about 500 AD the Dark Ages settled on Western Europe during which little of note is recorded. Military men continued to try to undermine defences and castles. One means was to drive a timbered heading under the wall and to set fire to the timbers in the hope that the heading would collapse, bringing the wall with it. Such effects are unintentionally achieved at times at the present day.

So for many years methods improved only gradually. In 1556 Georgius Agricola (the Latin form of his name — George Bauer) wrote *De re Metallica*. He is described in the Encyclopaedia Brittanica (1911) as the Father of Mineralogy and he was highly educated in medicine, philosophy, mathematics and other subjects. His deep studies in mineralogy and mining led him to publish another book in 1544 in which he laid down the first foundations of physical geology and criticised the theories of the ancients.

Woodcuts from his famous work are more often reproduced than the letterpress. They show the methods of mining, timbering, winding, pumping, ventilation, transporting and the methods of shaft sinking, and climbing out of them, as well as breaking rock by burning timber at the face. These methods had changed little before his time and continued in their basic forms until recent times. **Figure 1.**

Winding from deep pits was often done by *horse gins* or *whims*. One or two horses moving in a circle turned drums and, in some cases shown, they turned complex wooden wheeled contraptions, obviously the work of mill-wrights. The horse gin continued in use in many early railway tunnels, long after steam engines were in use. The simple jack-roll shown is used to this day in small works. It consists of a horizontal drum rotated by hand cranks and supported on easily erected timber supports. **Figure 2.**

When mines were in a suitable position water wheels were used to wind. In some cases tread-mills worked by men were used to turn the winding drum, this method was also used in some early types of crane. Timbered headings changed little before or after his time and will be discussed later.

According to Drinker "It was not until 1450 AD that the modern revival of tunnel construction on a large scale received its first impulse from Anne of Lusignan who, in that year, commenced a tunnel in the Alps between Nice and Genoa". Without further research this would appear to be a road tunnel. The Lusignans were a powerful French Crusading family who had become Kings of Cyprus but that line became extinct in 1475 AD so that may be why "the work was dropped". It was reopened in 1782 but after 2,500 metres had been completed it was abandoned when the French invaded in 1794.

The next critical step was the use of gunpowder. There is conflict over the dates of its first use in Europe between 1320 and 1380 AD. Curiously it seems to have been developed by various monks, mostly German.

In 1878 Drinker, who was a mining engineer, wrote — "Throughout the dark and Middle ages, war being the chief incentive to action, inventions were apt to take rise from warlike sources and thence to be subsequently applied to more peaceful pursuits". One hundred years later his words ring true with the Atomic Bomb and the rise of atomic power, both of which have triggered off an outburst of tunnelling.

Drinker reports that in 1397 a mine was blown as a military operation and powder was so used on a large scale in 1500 AD by Pedro Navaroo and by the Turks at Rhodes in 1529. He quotes Methesius — "The Great Generals always liked to carry miners with them. Thus in the siege of Vienna ... in order to find where the Turks were with their gallery, they are said to have placed a drum on the earth and at night when it reverberated, they knew where to sink down". This early form of geophysics is a reminder of the mining and counter-mining in World War I, so military intelligence is still to the fore.

FIGURE 1. *Use of horse-gin in mining. Agricola. 1556.*

FIGURE 2. *Overshot wheel used for raising water from mines. Agricola. 1556.*

Drinker gives a chronological Table of the development of explosive compounds, drilling and blasting which covers eight pages, starting with the Chinese in 80 AD and ending with dynamite in 1882.

The use of gunpowder in mines and tunnels seems to have been adopted slowly. Drilling and blasting in metal mines was said to have been invented in 1613 by Martin Weigel, though names are less important than the action. Small hand drills of steel were rotated while being struck with a hammer though this method probably dates back to Egypt. Wooden plugs were first used for tamping and later replaced by clay. The method was in regular use in Cornish mines by 1689 but safety fuse was not invented until 1831 by William Bickford.

The Spanish Conquistadores soon got busy mining in Mexico and it is there that the first North American tunnel is recorded, and described in a paper to the Institution of Civil Engineers by J.B. Body (1900), *The Drainage of the Valley of Mexico*. "The first systematic attempt to drain the waters of the valley was made in the year 1607, when Enrico Martinez drove a tunnel through the low depression of the mountains at Nochistongo, the northern extremity of the valley. This tunnel had a cross section measuring 13 feet 9 inches by 11 feet 6 inches and a length of 4.1 miles and was completed in the almost incredibly short time of 11 months; this, however, is explained by the fact that many thousands of Indians were employed throughout the whole of the time, working in a great number of shafts and that the soil offered no great resistance. Unfortunately the great urgency of the Spanish Government to get the work finished did not allow time to line the tunnel with masonry, and consequently, very soon after the first torrential rains passed through, the whole work collapsed, and the tunnel was left in this condition until the city was again inundated in the year 1629, continuing flooding for five consecutive years. The chronicles of the time record that 30,000 people perished during one month ...

"The project was again studied and in 1637 the project of converting the tunnel into an open cutting was accepted, but was not completed until the year 1789; the depth of this cutting is about 197 feet, the slopes being, on average, at an angle of 45 degrees ... The tunnel and cutting are estimated to have cost ... the lives of innumerable Indians who were forced to work as slaves. The bottom of the cutting being 29 feet 6 inches above the level of Lake Texcoco which is only three miles distant from the city gates, all this work proved practically useless as the danger of the lake rising and flooding the city during the seasons of heavy rain remained a nuisance to the inhabitants". This long quotation contains a lesson in every line and a warning to bureaucrats. The problems in Mexico City are being resolved today by extensive tunnelling.

Soft ground tunnelling during the Middle Ages has less history. Agricola in his woodcuts shows some form of timbered headings but the method became formalised once again, by Military Engineers though it is notable that these forward advances were the work of engineers and not of dashing cavalry officers.

This brings us inevitably to the famous French Marshal Vauban (1633–1707) who became famous by his design and construction of fortifications, but even more famous for his success in sieges and reducing other people's defences, sometimes of his own construction. Most of his work was in the Low Countries and consisted of digging deep ditches and building high earth banks so he was able to reveal to his disciples the lessons to be learned in soft ground which later evolved into Soil Mechanics. "He instituted a company of miners and the elaborate experiments carried out under his supervision resulted in the establishment of all the necessary formulae for military mining" (Encyc. Brit. 1911).

This mixing of the civil and military talents is shown in the foundation of the Ecole Polytechnique by the National Convention in France at the end of the 18th Century "où l'on instruit les jeunes gens, destinés à entrer dans les écoles spéciales d'artilleries, du génie, des mines, des ponts et des chaussées".

The teachings of Vauban were carried on at the British Military Seminary set up at Woolwich in 1719 which became the Royal Military Academy in 1741. In that year, John Muller was made Professor of Fortifications and Artillery and in 1757 wrote a book entitled *The Attack and Defence of Fortified Places*." In his third edition, 1770 he wrote — "The gallery being carried on some distance, carpenters follow the miners to prop the earth above, if it is not stiff clay or loam, to prevent it falling in. This is done by placing a piece of wood across the Gallery overhead, and putting a prop or upright at each end, fastened into the ground, then the earth is removed above this place, so as to slip a piece of deal board over it, which being afterwards supported in the same manner at the other end will sustain the earth."

This simplified recipe was extended by his successor, Isaac Landmann in a *Treatise on Mines* which has a plate showing in detail the method of what is known in England as a piled heading, while miners call the method spiling, and in the U.S.A., fore-poling.

The secret of the method in very loose ground is to ensure that more than the minimum of ground is never exposed and that it is quickly covered by timber. The face is close breasted and the sides and top advanced by the miner's labourer (and not as today, a carpenter) tapping forward the thin timber boards or piles supported by the frames of side-and head-trees, while the miner scratches away the ground and quickly supports it. In extreme cases considerable artistry was used, with short miniature headings to secure a small advance before opening up to larger spaces.

The journalist, Sandstrom described this fore-poling quite well, but went on to write "There probably exist throughout the Western world no men capable of driving a heading in this manner; it is a lost skill" (Sandstrom 1963). This is manifestly untrue but at the same time it is a reminder of how skills can be lost. But they can always be revived. It takes less than two years to train effective soldiers, and in the same time it has been proved possible to train simple men of many races in less than two years to carry out routine tunnelling work of different kinds, though the veteran is still needed in a crisis.

Chapter 2

RESULTS OF THE INDUSTRIAL REVOLUTION

THE CANAL AGE (1681–1840)

Human progress is apt to move by spasms or crises and the Canal Age had this effect on tunnelling. Civil engineering public works as we know them were scanty until impetus was imparted by the ambitions of Louis XIV (1638–1715), who with his Minister Colbert "striving to harness all industry and commerce to the service of the State" caused "the first modern tunnel constructed for transportation in the commercial sense of the word" (Drinker 1878). This was the Malpas tunnel, 515 feet long, 22 feet wide and 27 feet high completed in 1681. This is said to be the first blasted by gunpowder and forms part of the Languedoc Canal which Voltaire called "the most glorious monument of the reign of Louis XIV" (Derry & Trevor 1960), whose warlike ambitions led at the same time to further unwitting advances. Vauban was involved in the design of this canal with many staircases of locks, and also culverts to carry streams under the canal.

During the whole of this period, all over Europe a simmering of scientific advance was coming to the boil. The British Royal Society and the French Academy of Sciences led the field, while industrialists were making things happen with the aid of improving technologies.

For convenience we can equate the run up to the Canal Age and the Industrial Revolution, to the period from the English Civil War (1650) to the foundation of the Smeatonian Society of Civil Engineers in 1771. Trade and industry were expanding and especially the iron industry, not only for peaceful use but, as usual, for re-armament. It is curious to read that 12,000 to 18,000 cast iron cannon balls were expended every day in the two-month siege of Magdeburg in 1631.

Among the leaders of the new Iron Age were the Darbys — Abraham (1677–1717) and his son Abraham II (1711–1763). They proved that coal could be used instead of charcoal for smelting metals. This not only saved forests from devastation but caused an upsurge in coal mining with greater pressure on transport. Cast iron became a cheap and flexible material for innumerable purposes but also became a vital material in the coinciding advance of steam power, which eventually had an even greater effect on coal mining and led to fantastic advances all over the world.

The properties of steam were known to early philosophers. From 1700 when Savery produced his *Miners' Friend* or simple pump, through the advances of

Newcomen's engine (about 1712) the Cornish mines, as well as the coal fields, gained great benefit, leading up to James Watt's improving the reciprocating steam engine, by the addition of an external condenser. The early steam locomotives did not become a standard project until Stephenson's Rocket in 1829.

Throughout the ages Man had used the almost frictionless benefit of water for transport, in contrast to the intense difficulty of cartage over land. But rivers did not always flow in convenient directions. With the great demand for increased transport, canal building became a necessity. In many cases these were needed to cross hilly watersheds, so in order to avoid many flights of locks, a new type of tunnelling developed which had later repercussions.

The British Canal Age starts with the Duke of Bridgewater's canal for carrying coal direct from mines. The engineer was James Brindley and the first part was opened in 1761 before it became the Trent and Mersey Canal. This was of narrow gauge, taking boats up to 7 feet 4 inches in width. The first canal tunnel, Harecastle No 1 (driven between 1766 and 1777) was 2,880 yards long, 14 feet high but only 9 feet wide.

In 1827 a second Harecastle tunnel was completed by Telford this time 16 feet high and 14 feet wide with a tow-path, for horse and man were the only motive power until the railway age supervened. The last British canal tunnel was the Netherton on the Birmingham canal, 3,036 yards in length and 27 feet wide completed in 1858.

The importance of these tunnels is due to the way in which they led to a standardised method of tunnelling which spread around the world, known as the English System. In addition it led to the creation of a school of tunnel miners, distinct from the hordes of labourers, who had become known as *navvies* as they were often engaged on digging navigation channels. These tunnel miners were of necessity always on the move and they developed a style and tradition of their own, quite distinct from the coal miners who were tied to their pits. Thus for this period of over fifty years miners and engineers increased in numbers and accumulated experience which was invaluable in the creation of the Railway Age, which emerged in 1824 from primitive horse-drawn vehicles running on rails.

On the Continent, France was prolific in canals and their tunnels. These were of such sensible and generous dimensions that they were able to compete with the railways. Drinker in his Table 144 lists 24 French tunnels built between 1681 and 1840. Among much later ones the Royalcourt Tunnel on the Nord canal, completed in 1923, is 4,757 yards long, 33 feet wide and 28 feet three inches high. The largest navigable tunnel, the Rove on the Marseilles-Rhone Canal was opened in 1927 and is 72 feet wide with two towpaths and is 50 feet 6 inches from crown to invert (Pequignot 1963).

The tireless Drinker has compiled tables which occupy 79 pages of his large volume, in which he has set out details, under many headings, of 1,970 tunnels up

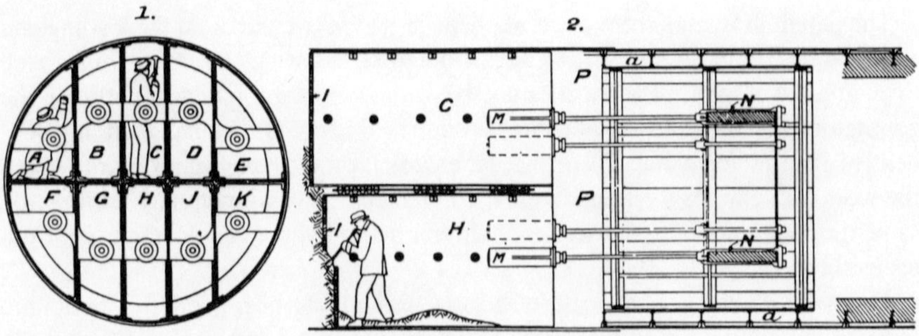

FIGURE 3. *Sir Marc Brunel's patent for a circular shield. (Copperthwaite)*

to the 1870's, these being mainly railway tunnels, including costs per lineal unit. Perhaps inflation was less rampant in his time. On canal tunnels data were more scarce but he gives the dimensions of 62 British tunnels of which only two, Harecastle II and Netherton come into the railway age.

THE THAMES TUNNEL, (1824–1842) AND THE BIRTH OF SHIELD TUNNELLING

No tunnel engineer can claim to be educated unless he has some knowledge of Sir Marc Isambard Brunel and his Thames Tunnel. It was the first shield driven, the first sub-aqueous and the first semi-mechanised tunnel lit by gas, and is still among the largest in cross section.

Sir Marc Brunel (1769–1849) was a French Royalist Naval Officer who escaped the guillotine in 1792 by going to New York where he practised as an Engineer and Architect. Then in 1799 he came to England, married Sophia Kingdom and soon was much in demand for his ingenious inventions. At the height of the Napoleonic wars he designed a machine for making much needed ships' blocks, which was made by Henry Maudsley who became the Father of Machine Tools.

Brunel worked for a time in Chatham Dockyard and had to drive a heading under one of the locks. From what followed there seem to be ample grounds for thinking that he imbibed his understanding of the true problems of soft ground tunnelling from the Chatham School of Military Engineering.

Brunel wrote "The great desideratum therefore consists in finding efficacious means of opening the ground in such a manner that no more earth shall be displaced than is to be filled by the shell or body of the tunnel and that the work shall be effected with certainty" (Copperthwaite 1906).

In 1818 Brunel took out British patent No 4204 for a circular iron tunnelling shield. Copperthwaite in his *Tunnelling Shields and the use of Compressed Air in Sub-aqueous Works* describes the specification in some detail, and added that it covers every subsequent development in the construction and working of tunnel shields. **Figure 3**.

The patent drawings show a circular iron shield in two parts. At the leading end are separate compartments side by side about three feet wide and six feet high, each to contain a miner. "Each cell may be forced forward independently of the contiguous one so that each workman is supposed to operate a small drift independently of the adjacent one. The front of the work is protected by small boards which the workman applies as he finds most convenient. Several men may work at the same time with perfect security, and without being liable to any obstruction from each other".

The specification goes on to describe how the cells can be propelled by hydraulic jacks in the main body of the shield which is then advanced by jacking from the following lining. "The body or shell of the tunnel may be made of brick or masonry but I prefer to make it of cast iron".

Thus Brunel invented the shield-driven, circular, cast-iron, segmental lined, tunnel and most important, the ability to line the tunnel at once as the shield advanced. He also invented the *tail* of the shield for he wrote — " ... I apply strong iron plates extending beyond the cell so as to overlap the shell of the tunnel previously made, thus protecting the space already cleared".

His patent also covered a form of tunnelling machine with rotating arms inside a circular shield, and followed by a spiral lining of masonry blocks. At that time no suitable motive power was available.

Soon after the Napoleonic war ended in 1815 the Thames below London Bridge and the newly built docks were cluttered with shipping on both banks so a scheme was put forward to drive a tunnel at Rotherhithe where 350 watermen ferried 3,700 passengers across daily.

In 1823 Brunel "brought before the public" his design for a shield for this tunnel, at a meeting of the Institution of Civil Engineers, of which he was a Member. This incorporated many of the ideas of his patent but it was designed for twin lanes of horsedrawn traffic with solid brick walls and brick arches over each lane. The shield consisted of twelve independent cells or units of cast iron each 22 feet 3 inches high and 3 feet 3 inches wide. So the shield was rectangular, 37 feet 6 inches wide. There were three floors in each cell or frame. Thus the shield was the equivalent of 36 simultaneous headings. The lining was of brickwork.

The tops of each frame and the sides of the outer ones carried iron plates propelled by screw jacks and acted as piles or fore-polings. In each frame the tunnel face was supported by horizontal boards 6 inches high and 3 inches thick, at each end of which were duplicate screw-jack supports. Alternate frames were advanced six at a time. **Figure 4**. The miners advanced the face one board at a time for 4½ inches under the protection of the roof and side plates. As each board was replaced it was supported by screw jacks from the adjacent frames so that when the face had been *taken down* from top to bottom the six frames were free to be jacked forward. The supports of the boards were then transferred to their own frame and

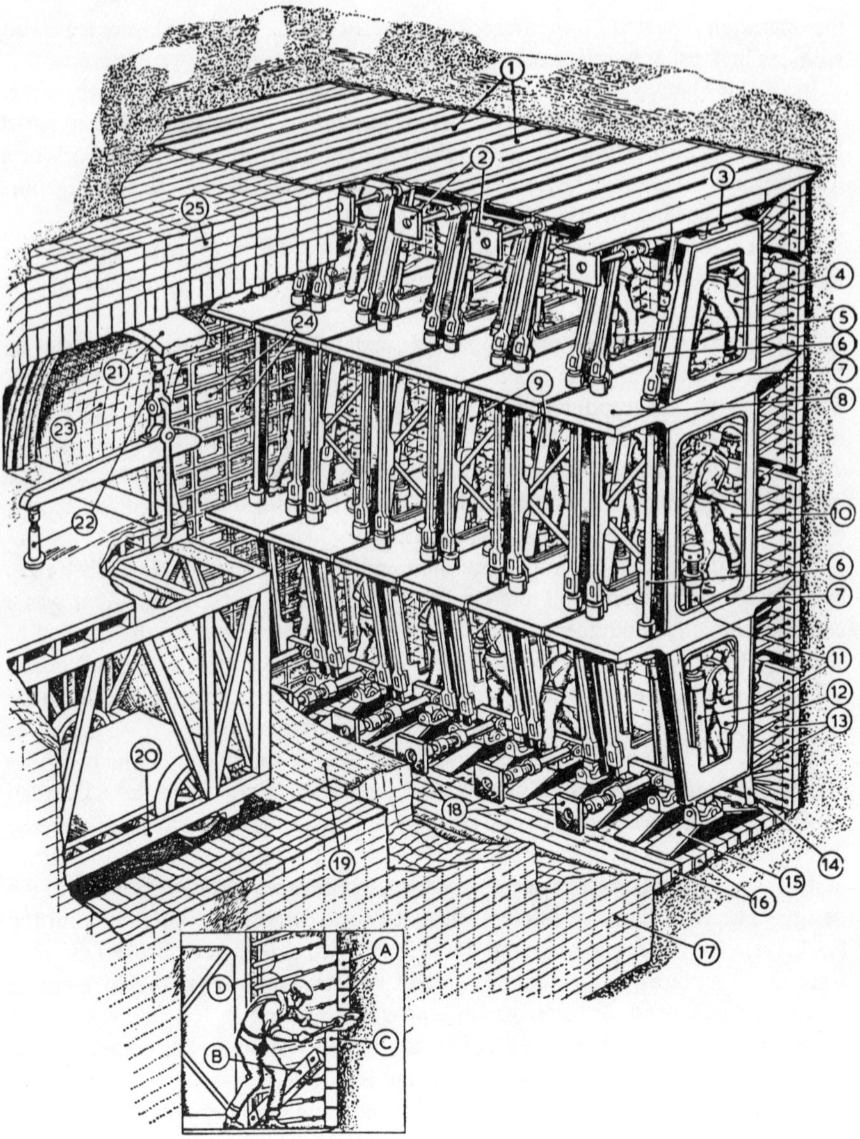

Key. 1: Top staves. 2: Top abutting screws. 3: Head. 4: Top box of frame no. 6. 5: Tail jack. 6: Wrought iron reinforcing member. 7: Cast iron side frame member. 8: Upper floor plate of frame no. 6. 9: Sling. 10: Middle box of frame no. 6. 11: Leg. 12: Bottom box of frame no. 6. 13: Poling boards. 14: Jack forcing down floor boards. 15: Shoe. 16: Floor boards on which brick roadways rest. 17: Brickwork of dividing wall. 18: Bottom abutting screws. 19: Brick roadway. 20: Travelling stage. 21: Roof centring. 22: Jacks for adjusting roof centring. 23: Western sidewall. 24: Side staves. 25: Roof brickwork.

A: Poling board moved forward. B: Poling board removed so that minor can excavate. C: Poling board that has not been moved forward. D: Poling screws.

the operation repeated on the frames previously static. The bricklayers followed with 4½ inch brick lining, working from a travelling stage.

In 1824 a Thames Tunnel Company was formed with Marc Brunel as engineer. The work was carried out by direct labour and his son Isambard was designated Resident Engineer at the age of 21, although he was really doing the work of a Contractor's Agent or Project Engineer, with direct control over the labour. (Pugsley & Harding 1976).

Brunel had paid £300 for a site investigation across the river but got little for his money as it seems to have been done by probing with a two inch diameter pipe. It gave the erroneous idea of clay the whole way.

A massive circular brick shaft of 50 feet diameter was sunk at Rotherhithe for 40 feet under its own weight and then to a depth of 65 feet by underpinning. **Figure 5**. The upper part was through river deposits and then in the Woolwich and Reading beds. These consist of thin layers of water-bearing sands and gravels, alternating with impervious material. The shaft was intended to be deeper but came to an end in the difficult Thanet Sands. Brunel built a large bricklined sump, with difficulty, by pumping from several wells simultaneously.

The shield was then built up in the shaft and launched on 28th November 1825 by jacking off the back wall and breaking through the brick lining. As it was a complete novelty, engineers and men had to train themselves in the method of working. Progress was 8 feet per week, working night and day in spite of water and soft ground in the invert. Brunel had allowed for a drainage heading leading to the steam pumps at the sump but the Directors refused to pay for it.

Among the lessons for Brunel (and ourselves) is the damage which can be caused by non-technical and opinionated Directors, for in June the Chairman insisted upon the men being put on piecework and for the shield to be advanced 9 inches at a time instead of the designed 4½ inches. This allowed ground to come in at the ends of the face boards and so upset the whole precautionary intention of the design. The men became careless on piecework, driven by pressure from the bricklayers for better progress and the shield became deranged. Broken castings were easily replaced as the patterns were kept. The shield was pulled together by chains and jacks and, in better ground, progress reached 14 feet in a week from Christmas till March 1827. The Directors allowed visitors in the tunnel to raise money at one shilling each but they were a nuisance and a risk.

Through March and April there were silt, sand and river deposits with constant outbursts of water through the gaps in the face boards. Forty men were needed to work the hand pumps to lift the water from the invert to run down troughs to the steam pumps at the shaft.

As a quantity of soft ground from the river bed had entered the shield, a compressed air diving bell, of the type designed by Smeaton, was borrowed from the West India Docks. Both Brunels descended in it and were able to pass a pipe from the bell into the shield. They were able to observe the type of cavity in the

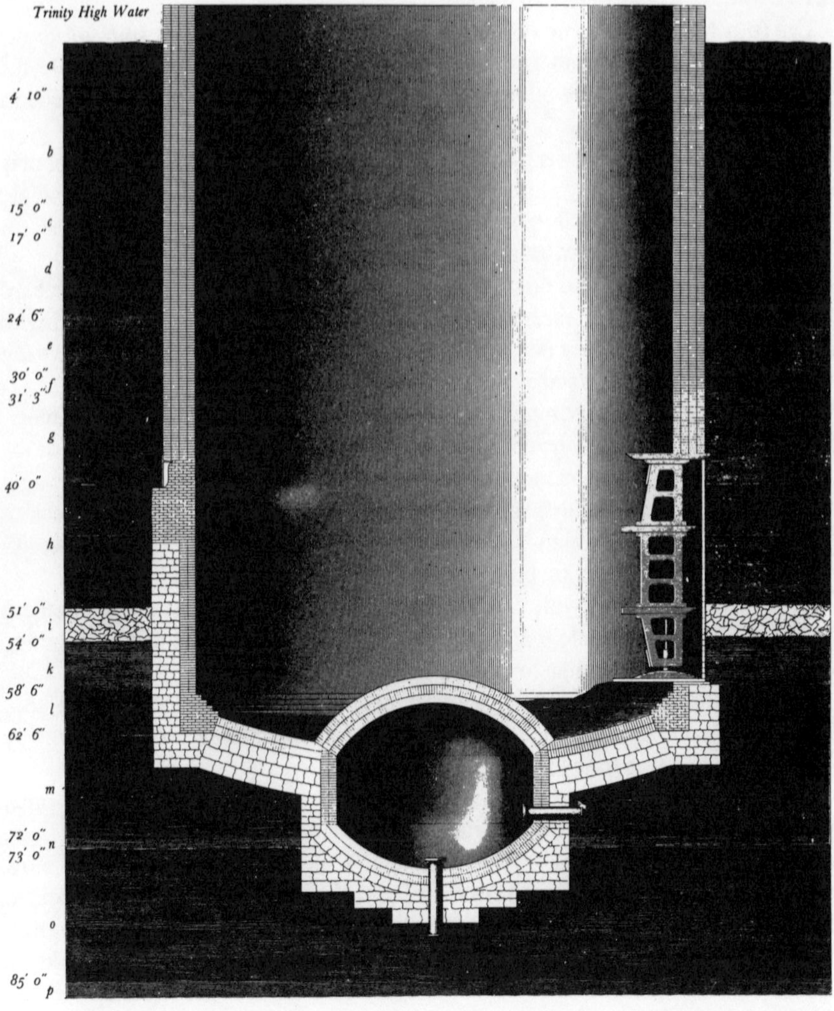

Key. a: Loam and made ground. b: Gravel, firm and free from water. c: Loose sand full of water. d: Very coarse gravel with large stones; "water so considerable over powered steam engine". e: "Quicksand" in flux equally abundant. f: Very base gravel; little water. g: Very stiff blue clay (London Clay). h: Clay more mottled in character (Woolwich and Reading beds); portion of silt; great number of shells. i: Chalk and clay combined; chisels to break out. k: Very firm green sand and gravel. l: Greener gravel and sand with chalk; red ochrous earth; a little water. m: Sandy loam quite free of water. n: Quicksand "copious in flux". o: Very green gravel; very little water. p: Light slate-coloured quicksand of very dangerous character; springs copious.

FIGURE 5. *Rotherhithe Shaft in Section. (Law)*

river bed and then organised the dumping of hundreds of bags of clay to fill the cavity. **Figure 6**.

On 18th May 1827 came the first irruptions of river water. In spite of heroic efforts by the miners, and Isambard, all had to make for the shaft while the whale-oil gas burners were put out. Isambard slid down an iron pipe and rescued a drowning pump driver.

The cavity in the river was filled by dumping a further 21,000 cubic feet of clay in bags amongst the busy shipping. The tunnel was pumped out and the shield found to be intact. Work started again and Isambard gave a dinner in the tunnel for forty guests, and for 120 workmen in the adjacent tunnel.

But when the tunnel was 605 feet long, on January 12th, 1828, while Isambard was examining the face, earth and water burst in and the tunnel again filled with water. Six men were drowned but Isambard was washed to safety, injured and unconscious, at the shaft stairway.

The tunnel was again pumped dry and the shield and brick lining found intact. But money had been exhausted, so a brick headwall was built at the back of the shield. Seven years were spent in raising money. The Duke of Wellington strongly supported the appeal and in a speech said, "Of my own knowledge I can speak of the interest excited in foreign Nations for the welfare and success of the undertaking: they look on it as the greatest work of Art ever contemplated".

Brunel was inundated with unhelpful suggestions which were useless. In these memorable words he wrote in his Diaries, "In every case they made the ground to suit the plan and not the plan to suit the ground" (Brunel M.).

Then on 5th December 1834 the Government made a first advance of £250,000, but by then young Brunel had become Engineer to the Great Western Railway which was a bad loss to the work.

Brunel's first act was to order a new and stronger shield built by Messrs Rennie to the same general design but weighing 140 tons in place of 80 tons. Henry Law, (1845) in a rare book *Memoirs of the Thames Tunnel* Part 1, gave a blow by blow weekly account of progress in the first half of the drive but Part 2 does not seem to have been written. Law gives detailed steel engravings of the second shield with all its ingenious intricacy, including a number of rocking arms which serve to hold the frames together but also allow for the independent forward movements. He also showed details of ingenious timbering during crises.

The second drive was of the same length but took twice as long, from a number of causes — some of them again due to the Chairman. The number of men was greatly reduced, gas was met with which caused sickness. Brunel was not allowed to sink the north shaft and to drive to meet the shield to give ventilation, access and increased pumping capacity. The loss of the driving power of his son, with his own failing health must have affected Brunel.

The tunnel was flooded three more times, but this seems to have become a matter

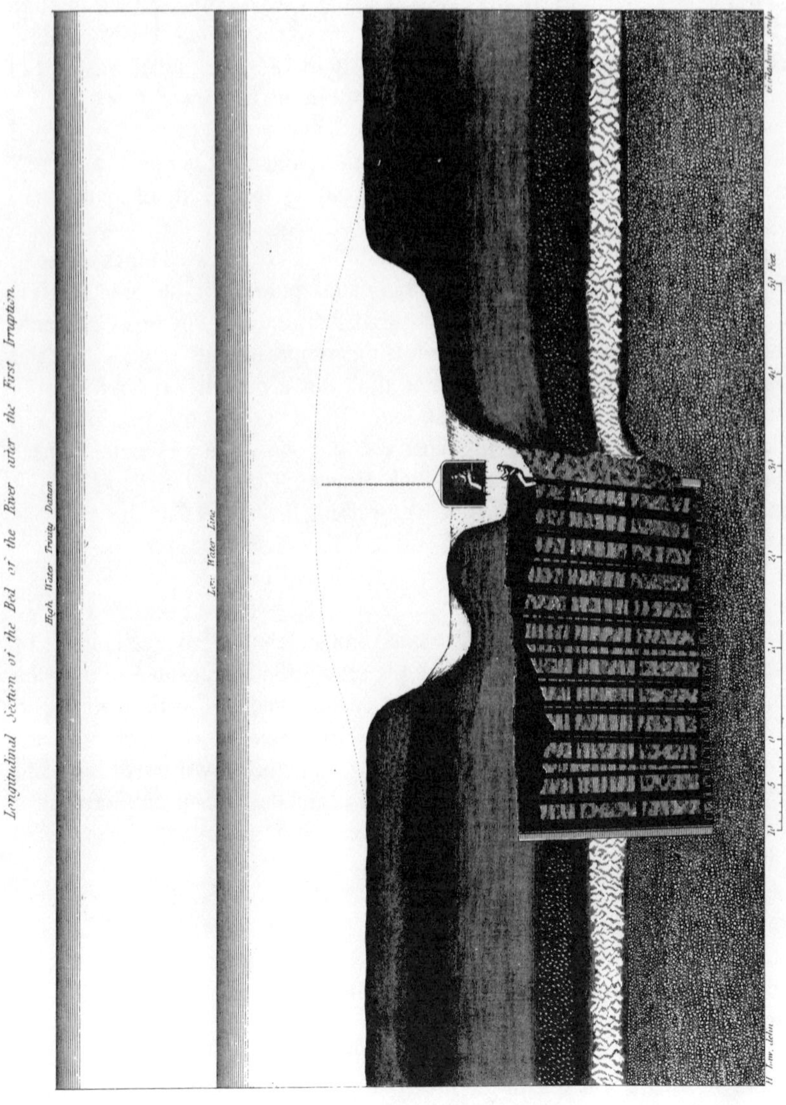

FIGURE 6. *Thames Tunnel — Inspection of river bed.*

of routine, coped with by clay bags dumped in the river. Eventually the north shaft was sunk, and a junction heading driven for sixty feet in August 1840. In honour of the occasion Brunel was knighted. Eventually the work was completed, and visited by Queen Victoria. Then in twenty-seven hours, 50,000 people walked through and in the next four months over a million visitors were counted. Unfortunately, the access roads were never made and eventually in 1865 the tunnel was sold to London's Underground railway system and is still in full use, exactly as built.

This shield tunnel was started 50 years before the next (the Tower Subway) and remained the first true sub-aqueous tunnel for even longer. It is remarkable how this heavy contraption was propelled day after day by the efforts of many men, working screw jacks with heavy spanners.

"Brunel's attention was called to the possibilities of compressed air as a help to keep the water out of his tunnel, but he did not attempt to put it into practice. It is evident, however, that his difficulties inspired Admiral Sir Thomas Cochrane, Lord Dundonald, to conceive a method of using compressed air in tunnelling and shaft sinking. In the year 1830 Cochrane took out a patent (No 6018) for an apparatus, which is in fact the modern air lock. The first time compressed air and the air lock were used in engineering work was in 1839 when a French engineer Triger sank an iron lined coal pit at Chalons, France. The word *air lock* was not used until 1851, when J. Hughes used it in describing the work of sinking caissons for the Rochester Bridge, England" (Hewett & Johannessen 1922).

THE COMING OF THE RAILWAYS (1824–1860)

Steam railways first started in Great Britain in 1824 and multiplied greatly in the 15 years between 1830 and 1845. This led to the revolution in transport all over the world and a great change in life-style. The canal tunnellers were available to exploit their skills in ever larger and more numerous tunnels. As other countries followed they developed their own systems to suit their own conditions and predilections. The Belgian railways started in 1835 and the French and Germans in 1837 while the Austrians followed in 1839. In the U.S.A. a number of lines started earlier, in about 1828.

Railway tunnelling dominated the period from 1824 to 1860 though aqueduct and canal tunnels continued, such as the New York-Croton (1837–42) and the Schuylkill-Susquehanna canal tunnel in Pennsylvania in 1818–21. Railway tunnelling continues to the present day but in differing and various forms while many other uses and needs have developed.

The Railway Age was dominated by the British Contractor, Thomas Brassey who died in 1870. He was one of the first to aim at improving relations between Engineer and Contractor. Even though there may be a gleam in the eye of the Engineer, giving him ideas, it is the Contractor who gives birth to the child of the Engineer's imagining. So such marital relationships are better if harmonious!

Brassey started as a contractor in 1835 as a man of education. He not only built a large share of British tunnels but also worked in most European countries as well as in Canada, India, Australia, and South America. His bodies of well-fed, well paid navvies and tunnellers set a standard which other countries had to follow.

In those days of intense private enterprise, in Britain all railway works had to obtain Parliamentary consent, which provided a livelihood for lawyers as well as consulting engineers as they argued their case before Committees.

In England many of the canal and railway tunnels were driven into hillsides of sedimentary deposits from most of the geological periods, generally requiring support. True, very hard rock tunnels were few. The tunnels were usually lined with brickwork with increasing skill. Only survivors who have worked with brickwork realise its virtues and versatility. When rock needed blasting by gunpowder the ground would be basically self-supporting, so random props would be used when safety demanded. But in loose or swelling ground heavy timbers were needed.

THE TUNNELLING SYSTEMS

Experience, passed on as it is gained, can lead to unofficial codes of practice. The methods learned in the Canal Age became the English System. As the railways spread there followed the Belgian, German and Austrian Systems. In France and the U.S.A. variations are recorded on all these systems, which were entirely dependent on timber for support before lining.

Until 1878,when Drinker ends his massive work, these were the basic methods of tunnelling and tunnel support. In his remorseless way he devotes 236 pages to detailed descriptions of many examples of all these systems with many drawings. He also drew up a long table comparing the pros and cons of the four main systems under headings. Segmental linings were then unknown. The main tools were the adze, axe, cross-cut saw and the timber auger.

a) English System

The English System of *tunnelling on bars* was also called length-work as a length of tunnel which might be, say, ten feet to twenty feet, would be dug and timbered in one operation and then lined. The length depended on the ground conditions and the length was taken out to full face. Today a few ancients survive who have seen the artistry of its performers as the method was still in use in large diameter sewers for London's main drainage till the 1930's. **Figure 7**.

The method is described in great detail by Simms in *Practical Tunnelling* (1844) from his own experience. First a bar-heading is driven, well ahead of each length. This would be piled or forepoled in loose ground or more simply timbered in good ground.

Holes for timber bars would have been left in the leading end of the previous

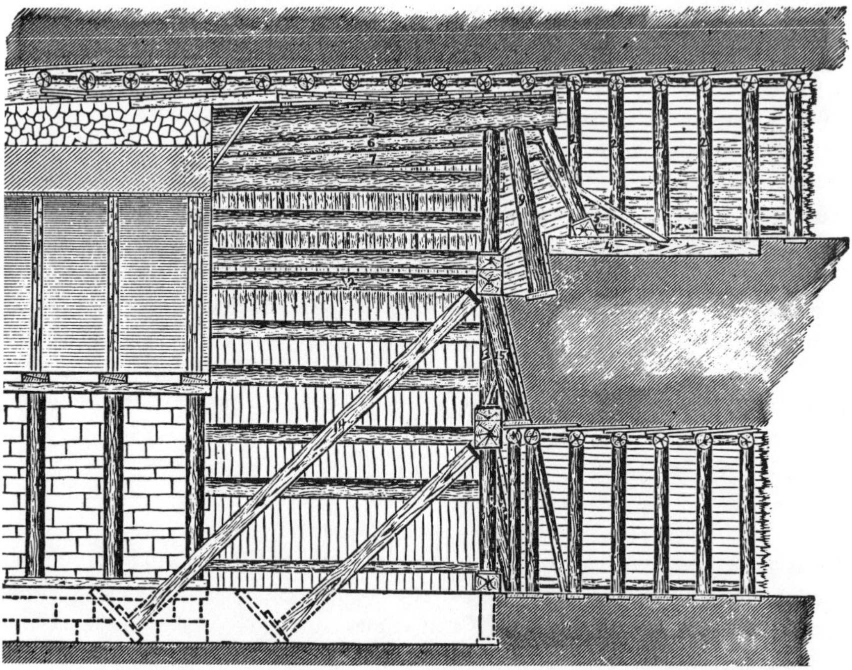

FIGURE 7. *Construction of Musconectong Tunnel. U.S.A. (Simms)*

bricked length. Two timber bars would be dragged and heaved into position, with the rear end in the bar-holes and the front ends supported by props in the heading. The miners would then work along each bar. The side trees and boards would be cut horizontally with the help of hand augers and timber slices. The miner's labourers would then drive short timber boards outwards and downwards over the bars while the miners scratched away the loose ground as the boards moved forward. Quantities of straw or long-litter (stable straw with its convenient content of soft material) would be at hand to stuff into any space which showed a tendency to run.

This would make room for the next pair of bars to be put in position with one end in the brick bar-hole and then propped at the leading end. This would be repeated until an arch was formed, with the last bars at axis level, acting then as walings. The face was supported by poling across the bar-props; on top of each would be placed a *Yankee Brob* consisting of a zed shaped piece of steel. The middle sat on the prop while one leg was against the end of the bar and the other on the inside of the top of the prop so that the bar acted as a stretcher to take the load carried by the vertical props.

The next step would be to place the Main Cill in front of the bar props and erect the inner line of main props on this, again to support the bars. It is then possible to dig out the remainder of the length as an open trench under the arch protection. The purpose of the main cill is to make it possible to underpin this as the digging proceeds. In a very high tunnel a second cill might be needed. For drainage, a subdrain was usually installed by digging out a small trench below the invert and laying an open jointed earthenware pipe surrounded by gravel. A chain would be left in the pipe and dragged forward for each length so that serious obstructions could be cleared out.

The length would then be handed over to the bricklayers while the gang would go to another face. While the bar-heading was being driven a few miners would work with the bricklayers, striking the bars as the brickwork rose and stacking them ready for the next length. The beauty of this system is the fact that only the poling boards are built in (in ground needing constant support). As the brickwork rises on timber *drums* and secures each setting of boards, the tops of the poling boards hold the toes of the row above so that the bars can be drawn safely. In good ground the timbering can be of a more open nature.

Soft woods were favoured for the bars, especially larch, as they creaked and groaned under stress before breaking, which could be forestalled by erecting horse-heads or other forms of support as required. The props and bars were used over and over again. The number of rings of brickwork was a matter for the Engineer's judgement.

Most of these tunnels were driven from a number of working shafts as in ancient times, but steam locos also needed ventilation, so some of the shafts were made bigger and permanent.

b) Belgian System

The Belgian system has been variously described. It consisted basically of bars supporting the top half followed by the lining of this half. The bottom would then be excavated while underpinning the completed arch above.

c) German System

In the German System usually two headings were driven at the bottom for each wall in which brickwork was built. A top heading was then used for widening out the upper section and the arch would be built upon the walls already in position.

d) Austrian System

The Austrian System was basically a full face excavation but with timber bars erected across and not along the tunnel, with rows of props. The bars were, in effect, tangents to the arch and supported poling boards placed along the length and not across the line of tunnel.

Tunnel Pressure on Timber

Drinker gives many striking drawings of the effect of swelling ground and heavy pressures on these timbered lengths. The advantage of the English System is that, as pressure increases, the timbering tightens (up to a point!). But the disadvantage of the length-work is the time it takes to excavate and line, in contrast to shield and segment work.

The London Clay, in which many tunnels have been made, is a stiff over-consolidated fissured clay and when it is first exposed it looks completely self-supporting to an innocent visitor. But if it is not quickly and permanently supported it can swell with remarkable results. Also large blocks can slide out of an unsupported face along fissures or *backs*.

Some British and Continental tunnels have shown signs of collapse after 100 years's service. This is often because, to save money, no brick invert had been built. This allowed the track to heave and the walls to move inwards.

Plant

During the Canal Age mechanical aids were almost nil, although crude steam pumps were being used, especially in mines. By 1824 steam cranes as well as locomotives were coming into use as well as better pumps, but even then for some years contractors had little capital and were men of small standing or education. It was the great Thomas Brassey who took up contracting in 1835 who changed the style and could meet the Stephensons and their fellow Engineers on the same level.

With a horse-gin, two horses would walk round a circle of, say, 30 feet in diameter, harnessed to an overhead bar. This would turn a wooden drum of, say, ten feet in diameter which rotated round a vertical shaft firmly fixed in the ground. Winding ropes were wound round the drum in opposite hand so that as one bucket rose the other descended. This might well keep pace with the rate of excavation and the lowering of supplies. The ropes passed over primitive head-gear.

During this time steam cranes and pumps were improving. The development of large displacement pumps was considered complete with the invention in 1840 by the American, Henry Rossiter Worthington, of the direct acting reciprocating pump, still highly popular today. William Mann patented compound air compressing or stage-pumping in 1829 but compressed air at high pressure was slow to develop until cooling was adopted in 1872. By then hydraulic power had reached a peak which it held until electric power entered the field.

Drilling was still by hand-held steels, which boys, called tool carriers, carried to and fro between the tunnel face and the blacksmiths' sharpening forges.

BRITISH RAILWAY TUNNELS

Canterbury-Whitstable

The early British rail tunnels set the pace and a few are constantly quoted as they gained publicity at the time and are better documented than others. But one of the earliest was little known. This was driven on behalf of George Stephenson between 1825 and 1830 on the Canterbury-Whitstable line. It was only twelve feet wide and of dubious brickwork. It was abandoned by British Rail in 1952 and later the University of Kent was built above it. A length of the abandoned tunnel collapsed in 1974 under one building which had to be demolished and the tunnel backfilled. A number of theories arose as to the cause, which was hastened by old age.

Kilsby Tunnnel

The Stephensons drove other early tunnels on the Liverpool-Manchester line and one notable tunnel is that at Kilsby, completed in 1838. It was 28 feet high, 26 feet wide and used 36 million bricks. The shafts included two for ventilation of 60 feet diameter, 120 and 90 feet deep. Eight months' delay occurred in overcoming water in the ground by an early form of Ground-Water Lowering by a series of pumps, worked simultaneously, by a series of bars connected to a steam engine.

Woodhead Tunnel

The original Woodhead tunnel linking the Sheffield-Manchester line under the Pennines was for single line working and 3 miles 13 yards long. The Engineer was Joseph Locke and it was driven in shaly rock and sandstone between 1838 and 1845. This proved a bottle neck so a second tunnel was driven alongside from 1847 to 1852.

The tunnel was lined with masonry and not brick and after 100 years the maintenance became so difficult that a new tunnel was driven for double lines from February 1949 to October 1953. So in spite of all modern conveniences the later tunnel took nearly as long as the original, which reflects credit on our predecessors.

Bletchingley and Saltwood Tunnels

Simms gives interesting details of the Bletchingley tunnel in swelling Weald Clay, and the Saltwood Tunnel, in lower Greensand. The trial shafts were sunk in 1840 and were only eleven feet in outside diameter with twelve inch thick brick lining, and were 100 feet deep. The tunnels were 30 feet wide and 30 feet high outside the 3 feet thick brick lining. The *lengths* were of 12 feet using larch bars. The bars and poling took up a space of about 18 inches beyond the brickwork.

Simms' descriptions give colour to the period. The winding in the shaft was done by horse-gin and he recorded at length the rate of winding of barrels of water to test the accuracy of the accepted figure for horse-power of 33,000 foot lb per minute.

His example of the costs for a twelve foot length at Bletchingley paint the picture which could apply to most rail tunnels of the period.

Drive the top heading 4 yards	@	15/-		£ 3 0 0
4 men getting in tops 4 shifts	@	6/-		£ 4 16 0
Remainder (say 8 shifts)				
4 Miners	@	6/-	£1 4 0	
1 Mongrel	@	4/6	4 6	
3 Fillers	@	3/6	10 6	
4 Banksmen	@	3/6	14 0	
1 Hooker-on	@	3/-	3 0	
2 Horses and Driver	@	7/-	14 0	
			£3 10 0	
Multiply by eight shifts				£28 0 0
Powder and candles				3 0 0
Tools, etc				1 5 0
Superintendence 8 shifts	@	7/-		2 16 0
				£42 17 0

or £10 14 3 per linear yard.

The brickwork was carried out by 4 bricklayers and 7 labourers. For an average over 20 lengths —

Setting ground moulds for springing	2 days
Springing to arch and keying-in	2½ days
per lengths	4½ days

Box Tunnel

When Isambard Kingdom Brunel became Engineer to the new Great Western Railway in 1833 at the age of 27 the key to his critical path between London and Bristol was the Box Tunnel under the Cotswold Hills near Corsham.

I.K. Brunel had gained a world-wide reputation by his work and adventures on his father's Thames Tunnel which is an interlude on its own. This may be why the Box Tunnel obtained more publicity than those of his contemporaries and more protests from arm-chair and Clubland Amateurs on its impossibility. More has been written on its human than its engineering aspects.

The Box Tunnel (1837-41) is 1 mile 321 yards long at a down grade from east to west of 1 in 100. It starts in the Greater Oolite, a Jurassic limestone famous as Bath Stone. This was unlined with a cathedral-like arch above, and gunpowder was used. Years later a protective roof had to be placed. For many years Bath stone has been quarried by underground workings at Corsham. The tunnel then enters Fullers earth, needing brick lining; then in the Inferior Oolite which is only inferior because it is under the Greater. The last part is in Lias Clay. **Figures 8 and 9**.

Brunel (to his credit) wrote to the Company Directors (Brunel I.K.) to say that

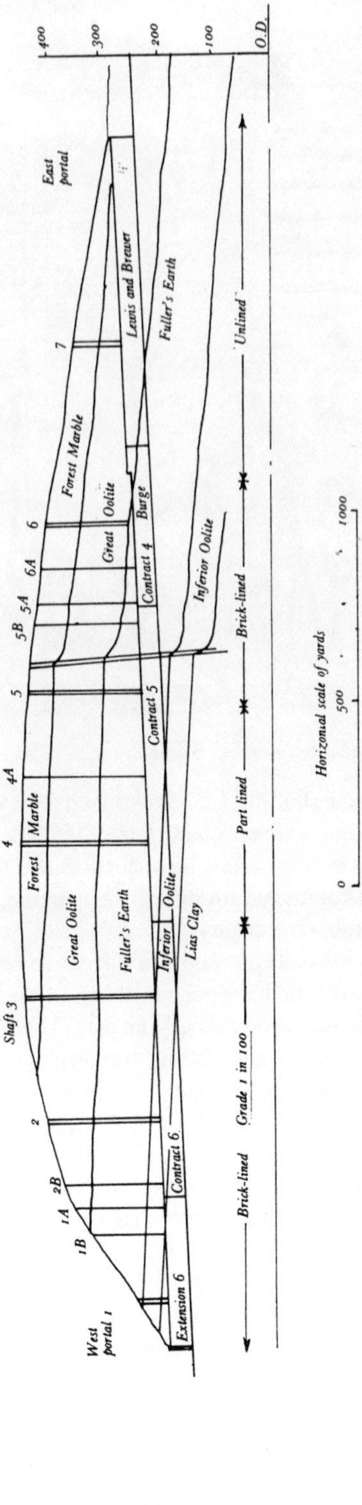

FIGURE 8. *Box Tunnel—Longitudinal section.*

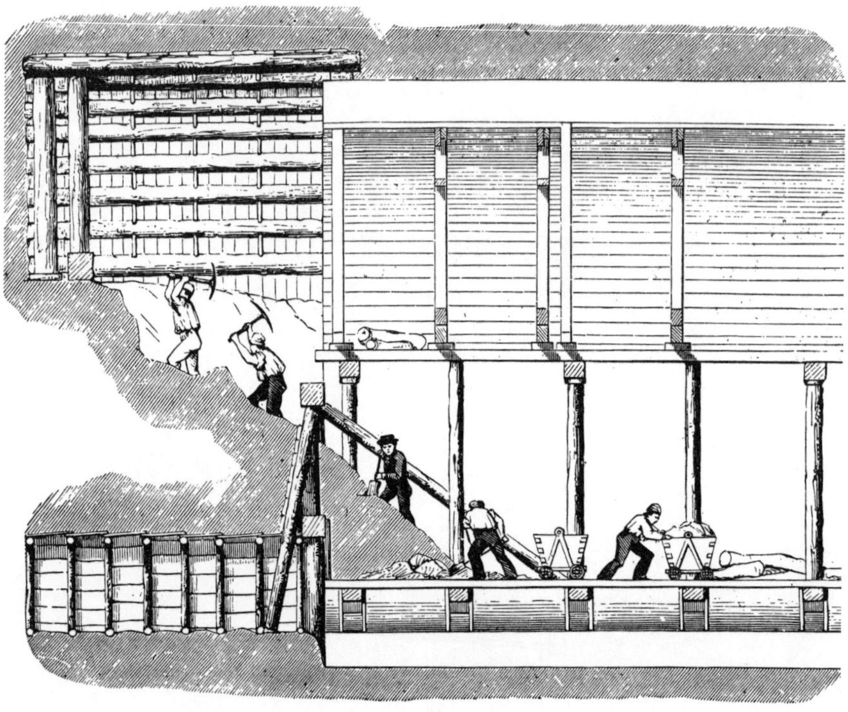

FIGURE 9. *Box Tunnel — Tunnelling method. (Simms)*

he intended to finish all the shafts before drawing up the contracts for the tunnel, so that "the materials through which each portion of the latter is to be carried may be ascertained and worked by the parties most accustomed to the particular description of material and not contracted for blindly on a mere speculation". No wonder that his admirers outnumber his critics.

Brunel had generous ideas. If his gauge of 7 feet had been adopted in place of 4 feet 8½ inches, railways would have been faster and more commodious, so lasting longer commercially, though at greater capital cost. The shafts at Box were 25 feet in diameter, 7 in number, the deepest being 290 feet.

Although the Brunels had used the latest steam plant and winding gear on the Thames Tunnel, the winding at Box was all by horse-gins as the choice of the contractor. It may have been cheaper to use horses to wind, than to use them to haul coal for engines from Somerset over atrocious or non-existent roads. This helps to emphasise the difference which railways made to transport as each length was completed. **Figure 10**.

The peak force consisted of 4,000 men and 300 horses. 30 million bricks were carted to the site by 100 horses and a ton of candles used every week. Simms

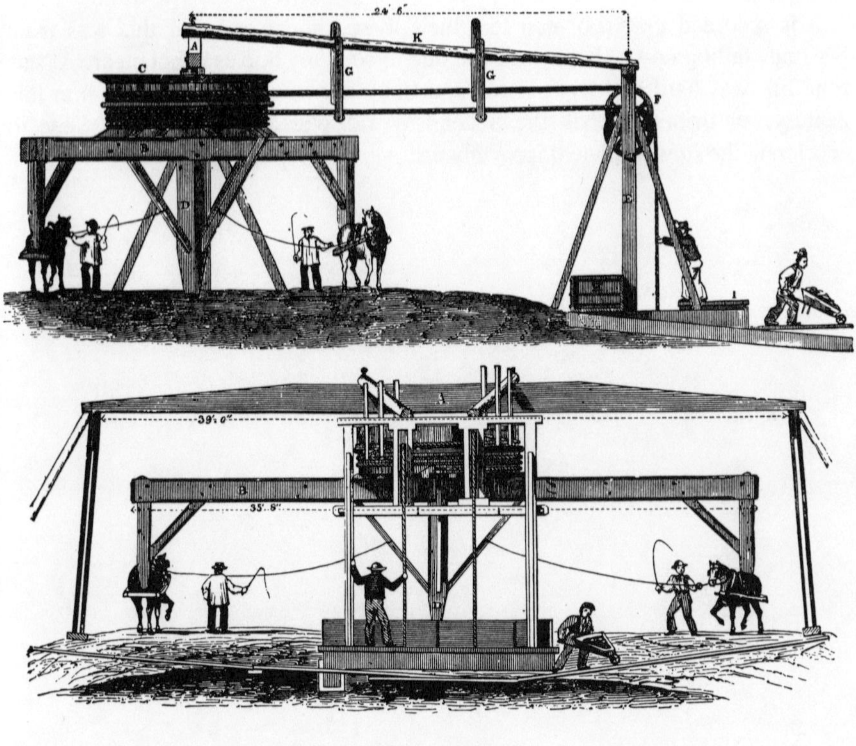

FIGURE 10. *Box Tunnel — Horse gins for winding. (Simms)*

records that "work was suspended from November 1837 to July 1838 as water from fissures gained on the steam pumps employed, reaching 50 feet up the shaft. A second pump of 50 hp was applied and water pumped at the rate of 32,000 hogsheads per day". A hogshead of beer is said to contain 54 gallons and seems a convenient measure.

The Box contract was engrossed on parchment by scribes with all the contemporary flourishes and curlicues of penmanship. It covered several 2 feet wide sheets in deep legal jargon with no punctuation or paragraphs. In contrast the specification was terse and to the point. There was only one item for driving the tunnel and six for various linings but only with unit prices and no quantities.

The contract drawings were prints from steel engravings and hand coloured. There was one longitudinal skeleton section and several cross sections with thick linings above and thin below or vice-versa. Presumably this was to suit the changes, as the tunnel passed gradually from hard to soft and then back to hard ground. As there were no quantities the measurement would follow the course of the ground as work proceeded.

It is recorded that 100 men lost their lives, though whether this was from blasting, falling down shafts, or even due to so many horses is not clear. At this time life was hard and lives cheap and this harshness continued well into this century. In Britain it took the Second World War and the Welfare State to accelerate the slowly made improvements.

Chapter 3

A CRITICAL DECADE (1860–1870)

This decade, a mixture of war and peace, marked a turning point in world history and so affected tunnelling. The end of the Civil War (1865) in the United States led to the drive to *go West* to open up the continent, with all that this meant to tunnelling and mining. In Europe the ambitions of Napoleon III and the cunning of Bismark led to several wars which ended with the Franco-Prussian War in 1870. This led to the formation of the German Empire and the release of the stored-up talents of the German people. Italy became a nation. The British and Germans began to invest in South America. In Mexico, once Emperor Maximilian had been shot, President Diaz began his rise to power and the modernising of Mexico. Japan, having defeated their ultra-isolationists in civil war, decided to *go Western* and invited British Engineers to build their first railway in 1870. The European Powers began to divide up Africa.

During these turbulent events significant changes took place in the whole tunnelling picture.

The overcrowding and deplorable conditions had led to the first stage of Sir Joseph Bazalgette's London Main Drainage (1855-1874). This included the High Level, Middle Level and Low Level main sewers on either bank of the River Thames. These were up to twelve feet in diameter or over, tunnelled "on bars" and lined with high grade brickwork. This wiped out water borne diseases, typhoid and cholera. **Figure 11**.

It is hard to over-emphasise the vital importance of the railways in revolutionising Social History during the following period of peaceful expansion, although some railways were ominously described as *strategic*, which later proved only too true.

The congested streets brought about the first city *underground railway* or *Metro*, the London Inner Circle, mostly in cut and cover but with tunnels at Farringdon Street and Campden Hill. This started in 1864 and went on expanding for years.

The revival of shield tunnelling by Greathead's Tower Subway which followed Barlow's patent of 1864 led after seventeen years to the start of the London shield driven tubes. This immediately set off the outburst of sub-aqueous railway tunnels. It was some years before other cities followed the lead in building subways and Metros.

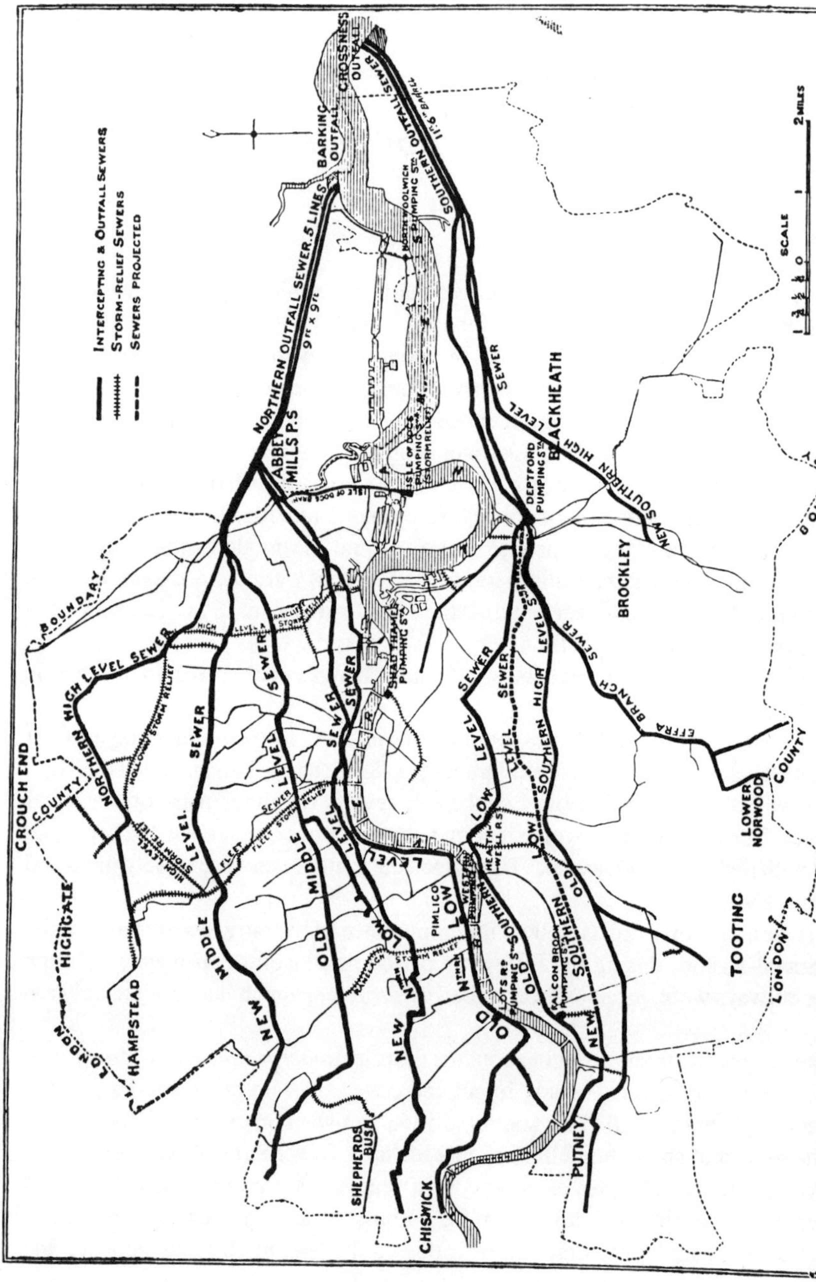

FIGURE 11. *London Main Drainage—Bazalgette's scheme. (Eng. Wonders)*

Drinker in his Table 21 describes all U.S.A. patents for rock drills from 1849 to 1880 covering 14 pages listing 187 patents. In Table 25 he lists 130 British patents. Full face tunnelling machines were patented by H.N. Penrice in 1856 and F.E.B. Beaumont of Channel Tunnel fame in 1864 and 1875. Patents for explosives were also numerous. This gives an idea of the ferment of activity in this prolific period. One name is usually connected with an invention but what Dr Johnson said to James Boswell seems appropriate (Boswell 1950). He said that "human experience which was constantly contradicting theory was the great test of truth. Experience built upon the discoveries of a great many minds was always of greater weight than the mere working of one mind, which can do little of itself".

Until this period mountain ranges seemed too expensive to attack, but as commercial pressures to cross them mounted, fresh inventions gave fresh ideas. In 1863 Nobel invented nitro-glycerine and in 1868, dynamite. Air compressors were improving and mechanical drills were beginning to appear. Bessemer in 1856 invented his Converter so, from then on, mild steel was becoming an article of commerce with wide spreading results.

In 1857 the first of the historic series of Alpine tunnels started under the Col de Fréjus near Mont Cenis. Hand drilling was used for the first three years until Sommeiller produced his rotary-percussion drilling machine, driven by compressed air, in 1861. Progress increased eventually to five times that of the hand drilling length, It was opened in 1871, 14,052 yards long to link the French and Italian railways.

The decade ended with the States of America firmly, if reluctantly, United. The far from united States of Europe embarked upon a period of over forty years of peaceful co-existence. Civil Engineers and their men had the courage to tackle more formidable obstacles than ever before once they had been able to inspire courage into the financiers to spend money more adventurously. Railways formed the bulk of fresh tunnelling as it was the only form of general rapid transit until after the first World War.

Parliament conferred the title of Empress of India on Queen Victoria in 1876 and steam navigation was expanding rapidly which brought development to far places. Steam remained the great and almost only motive power. Figure 26 shows the relation in time of the start of the many different forms of activity and the main scientific and technical advances which brought them about or helped their advance.

Most great cities are built beside large rivers or estuaries and consequently on soft ground. As they grew they needed more types of tunnels so that *schools* of soft ground tunnelling developed in them, as did *schools* of Painting in the Renaissance.

These cities needed water supply and electric power which had a dual effect. Aqueducts, one of the earliest forms of tunnel, produced, for example, a shield-driven tunnel under the Mersey while electric power stations needed cooling water and cable tunnels in soft ground. But paradoxically both these needs led to an extension of rock tunnelling. Dams, with their own quota of tunnels, in hills and mountains needed rock tunnels to carry their water to the cities while hydro-electric power became more advanced. This from its nature produced a fresh outburst of rock tunnelling.

More can be learned from the first outcome of a new advance than subsequent repetitions, these become too many for individual attention and some have to be treated in more general fashion. Many contrasting classes of work run concurrently.

Variations in the ground complicate the story. Ground can be either dry or waterbearing both under land or under water. Two new expedients were introduced, the Greathead shield (a simplification of Brunel's patent) and the application of Lord Dundonald's patent of the use of compressed air to hold back water. So historic tunnels were driven first without a shield or compressed air, without a shield and with compressed air, with a shield and no air and eventually with both a shield and compressed air. There are five types of new works which ushered in this new age of tunnelling, these are treated in separate sections.

Chapter 4

THE GREAT ADVANCE (1870–1914)

THE FIRST ALPINE RAILWAY TUNNELS (1857–1906)

As the railways spread their networks they were faced with two major challenges. One was the need to dive under large rivers often in soft, water-logged sedimentary deposits. In that case it was possible to carry out some form of site investigation with the rudimentary methods of the time. The second challenge came from vast mountain ranges such as the Swiss Alps whose whole existence stemmed from the extraordinary contortions forced by earth movements on a gigantic scale. Tunnels of great length would have to be driven into hard distorted rock where deep borings were out of the question. Reliance would have to be placed on the Horoscopes cast by the comparatively new profession of Geologist.

History was made by four major rail tunnels through the Alps where historic passes showed the need. To these must be added the first mountain tunnel in the U.S.A. — the Hoosac. Much has been learned from each in turn. The first Alpine tunnel was the Mont Cenis or Fréjus, proposed in 1838 and eventually started in 1857. It took 13 years for its length of 40,000 feet. The Hoosac tunnel, mooted in 1826 as a canal tunnel started in 1856 and took 21 years in a series of *stop-go* spasms for a length of 20,000 feet. So both tunnels suffered long periods of gestation before being carried out within the limitations of the age.

Mont Cenis, or Fréjus Tunnel (1857–1870)

The Mont Cenis tunnel, driven from both ends in metamorphic sediments, greatly disturbed, is 7½ miles long and joins the French to Italian railways. At first slow progress was made by hand drilling and ventilation was bad. Before this many brains were working on possible developments of mechanical drilling machines. Sommeiller's compressed air drill was decided upon by the Official Commission for use in the tunnel in 1857 but this did not come into action until the start of 1861. In 1852 the Swiss physicist Colladon had proposed the use of water power to drive air compressors. The combination of these two inventions increased the rate of progress by five times, after hand drilling was abandoned. The drilling machines combined impact, rotation and automatic feed. Water power, to drive air compressors was used in all the later Alpine tunnels. Gunpowder was the explosive. The Belgian system was used and the brick lining left an inside dimension of 26 feet 3 inches wide and 24 feet 7 inches high for double tracks. Many accounts are available including a paper by Sopwith (I.C.E. Vol XXIII) in which he considered that the cost of providing and running the mechanical plant and the cost of the outside works to harness the water power balanced the cost of the greatly increased progress. There were 28 fatal accidents. Sandstrom (1963) writes "At the opening celebration toasts were drunk in torrents of champagne but none to the memory of Sommeiller. He was not present, having died of overwork in the fashion of all great tunnelling men".

The Hoosac Tunnel (1856–1864–1875)

The Hoosac tunnel to link Boston and the Hudson River valley was driven in less mountainous conditions, in mica schist and granite for four and three quarter miles. It started slowly by hand drilling while many tests of drilling machines were being made under State supervision. The contractor left in 1861 after a quarrel on payment. After studying the European tunnels the State of Massachussettes took over by direct labour and had a dam made to provide water power for compressors at the east portal. Steam was used at the west end and eventually a central shaft was sunk. After many changes of engineer and reverting to contracts, drilling got well under way using Burleigh, Ingersoll and Wood drills. A bottom heading was driven to help drainage, a modified English system was used and the tunnel lined with brick and masonry. In places decomposed rock caused swelling pressures. Dynamite was tried but abandoned in preference to Nitro-glycerine and mica powder. Nobel had patented this in 1865 and dynamite in 1867. A timber carriage was used to help in removing spoil, a first trend towards mechanisation.

Saint Gothard (1872–1882)

The success of Mont Cenis so revolutionised traffic between France and Britain into Italy and its ports (on the way to India) that the Swiss determined to press on with their long considered project for a tunnel under the St Gothard Pass. The approaches include seven corkscrew tunnels which are a feature of other Swiss railways. As the track climbs the mountain side it disappears into a tunnel and, after making a complete circle, emerges many feet higher, above the point of entry.

The St Gothard tunnel, 9½ miles long, is driven in granite, gneiss, mica schist and Jurassic limestone. By this time better drilling machines were available and those of Ferraux, Ferrari and McKean were used. They were mounted upon rail carriages, six or eight drills at a time, and powered by double-acting water-cooled air compressors at about 100 lb per square inch pressure, driven as before by water power. The Belgian system was used. During 1880 disintegrated felspar produced great swelling pressures which threatened to crush every attempt at lining. Granite arches five feet thick supported by seven feet thick granite walls suffered crushing.

On the South side at one time 6,000 gallons of water per minute had to be dealt with but presumably this eventually drained away. Dynamite was chosen for blasting. The camp conditions were bad. One hundred and seventy-seven men are said to have died in the tunnel out of a total of 300. The contractor, Le Favre, died of apoplexy in the tunnel and there is a story that when the tunnels junctioned his portrait was first passed through. Debris was loaded into baskets for carrying along the line of trucks to be filled. In the contemporary Misconetcong Tunnel in the U.S.A. (1872) a form of timber gantry was used to allow trucks from the upper bench to pass over those at invert level to speed up loading.

The Arlberg Tunnel (1880–1884)
The Arlberg Tunnel is 6½ miles long and joins Austria with Switzerland. It went very smoothly by profiting from the previous experience. An entirely new form of drill was used, the Brandt, which was driven and advanced by hydraulic power and not air. (This type is now back in favour). Water was met with at the west end in micaceous fissures and a bottom heading was driven. *Break-ups* every 79 feet at the east and 216 feet at the west allowed a number of faces to be worked at the same time. Better ventilation was provided through large pipes and water jets were used to lay the dust after blasting by dynamite. The effect of all these improvements is shown by the average progress — Mont Cenis 2.57 yards per day, St Gothard 6.01 yards and Arlberg 9.07 yards per day (Eng. Wonders of the World 1909).

The Simplon Tunnel (1898–1906)
The Simplon tunnel, 12½ miles long under the Simplon Pass emerges into the start of the modern age when electric power and light became available as well as mild steel and other benefits. The progress on average was twice that at Arlberg, inspite of difficulties. Twin single line tunnels were chosen which greatly helped construction and future maintenance. Two parallel headings were driven up hill from both portals with the intention to break up to roof level and drive top headings both ways for enlarging ready for the masons to line. After a time the top gallery was abandoned and the tunnel enlarged from the bottom heading to give easier ventilation. Cross passages between the headings were kept closed except for the leading one so that air, being pumped up fifteen inch diameter pipes to one face could return by the other. Cooling water sprays were provided as routine.

Brandt hydraulic drills were again used and carried three at a time on a carriage anchored to the walls by hydraulic jacks. Better techniques had been learned for blasting by dynamite while mild steel was used as flooring to assist in shovelling debris. The ground was mostly gneiss, mica schist and limestone. Timber was the principal means of support, as in all these tunnels. Very high temperatures and pressures were met with at 7,000 feet under the mountain. At 2¾ miles from the Italian portal quantities of cold water at high pressure coincided with intense squeezing by decomposed schist which crushed timbers and closed the heading. The tunnel had to be enlarged in three stages each with formidable lining problems. Frames made of 16 inch by 6 inch rolled steel joists supporting 20 inch square pitch pine timbers still needed filling in with quick setting concrete to hold the ground.

As the tunnel faces neared their junction very hot water was met with in both faces of Gallery No 1. Fortunately gallery No 2 proved to be in sound rock and was driven past this point to connect with Gallery No 1, while the water drained away. The last 245 metres took six months to drive.

The Lotschberg Tunnel (1906–1913)

The Lotschberg tunnel, 8½ miles long is entirely in Switzerland and consequently seems to be treated by writers rather separately from the international links. By this time procedures had become standardised, leading to over confidence. The warning of a possible infilled valley was ignored. When a round was fired in the double track tunnel, boulders, silt, mud and sand ran 4,300 feet into the tunnel, burying a gang of twenty-five men. The tunnel was diverted, adding half a mile to its length and was successfully concluded.

The Andes Tunnel (1905–1906)

The Andes tunnel on the Trans-Andine Railway between Argentina and Chile, 2 miles long was at an elevation of 10,420 feet compared with the Simplon at 2,313 feet and Mont Cenis at 4,245 feet above sea level. It was started locally using hand labour in drilling and after three years' work only 327 metres had been bored. Then Messrs Walker, (of the Severn Tunnel) took over and introduced drilling machines which increased progress from 9 metres to 45 metres per month. The tunnel was mostly in rock and was lined with portland cement concrete. The air compressors and lighting generators were driven by diesel engines of 80 and 120 horse power. Owing to the altitude they only developed 70 per cent of their normal power to the surprise of some of the engineers (Fawcett B. 1906).

SUB-AQUEOUS TUNNELS BEFORE THE SHIELD AGE (1870–1886)

a) In Free Air

From 1870 onwards a number of under water tunnels without the protection of a shield were carried out or attempted. In 1870 several 5 feet diameter brick lined tunnels were driven as much as two miles out under Lake Michigan, to provide (what was then) fresh water for Chicago. These were in the local soft clay, though some water was met with. Inflammable gas was met in cavities and was ignited as soon as it began to escape. "At such times the miners fell with their faces to the ground and thus escaped without any greater injury than singed beards and eyelashes" (Drinker 1878). Never a dull moment!

The intake shafts were sunk inside large timber cribs 40 feet high with five sides each 55 feet long. The cribs were built horizontally, tipped by jacks during launching to float vertically, towed to the site and sunk by filling with rubble. This forestalls the oil platforms in the North Sea.

Detroit River Railway Tunnel (1872–1873)

Borings were made with a wood auger 1 ¼ inches in diameter. This was drawn up, usually once in every ten feet, and it usually brought up the whole strata! Twelve feet of stiff clay was expected above the tunnels which were to be twin, brick lined. The ground turned out to be clay but with layers of sands and gravel and also contained gas in places. After driving had been carried on in the old fashioned way, without a shield and after many irruptions of water the contractors became very much discouraged. They asked to be relieved of their contract as the work was costing four times more than their price. After 1,000 feet had been driven the work was abandoned.

In a paper to the American Society of Civil Engineers in 1872 the Engineer, Mr Cheeseborough gave reasons why the work could not have been carried out by the pneumatic process. "Besides the fact that no horizontal drift is known to have been made in this manner, it will be sufficient to state, to those familiar with the process, that work excavated under 90 to 100 feet head of water is not only very expensive but hazardous to human life". Six years later the method was brought into use elsewhere.

The Mersey Railway Tunnel (1879–1886)

This double line rail tunnel was driven under the River Mersey to connect Liverpool with Birkenhead mostly through red sandstone, without explosives and with few difficulties. Drainage shafts were sunk on either bank and drainage tunnels driven to the lowest part of the tunnel. The tunnel was lined with six courses of brickwork under the river with four courses elsewhere. The Encyclopaedia Britannica states that a Beaumont tunnelling machine was used on the drainage tunnels of 7 feet 4 inches diameter.

Only one fault was found in the sandstone. Beam engines were chosen for the pumps for reliability, capable of dealing with 7,000 to 8,000 gallons per minute. The level section was 145 feet below high water and at one time twenty four faces were at work. Also boreholes had proved the existence of the sandstone satisfactorily.

The Severn Tunnel (1872–1885)

This tunnel to take the Great Western Railway into Wales was driven in Trias and Coal Measures of which most was sandstone. The river is two miles wide but the total length is 4½ miles with a ruling grade of 1 in 90 in order to get below a buried channel, called the Shoots. The tunnelling itself was straightforward when actually at work, on the English System from a number of break-ups fed by a heading driven at invert level.

It seems that the work started by direct labour as one account states that the directors, after some shafts had been sunk, decided to let the tunnel contract, but on second thoughts determined to first *prove the ground* by a heading through the whole length of the tunnel. From one shaft on the west bank a drainage heading had been driven so as to drain the lowest part of the tunnel, as on the Mersey.

But while driving uphill on the main tunnel grade an inrush of water occurred, now known as the Great Spring. The men were able to escape as the filling of the drainage tunnel gave them time. This led to the Engineer, Sir John Hawkshaw, demanding that the contract be given to Mr T.A. Walker who had much experience on the London Underground. The contract was signed at the end of 1879. The first work had started tentatively in 1872.

Fresh shafts were sunk and much ingenious work done by divers in clearing the drainage tunnel. Trouble occurred with the pumps, driven by beam engines. The pump rods consisted of baulks of timber 15 inches square and 45 feet long. The rising main was made of 40 inch wrought iron pipes. The Great Spring is still delivering 27 million gallons a day to local water supplies.

The tunnel was flooded again when a high wave in the Severn topped the banks and again from a water burst on the eastern side. But when the tunnel was not under water, work went on in the fashion of the time, being timbered all the way with some use of explosives. The line was opened in 1886.

The Channel Tunnel (Part one: 1882)

In 1872 the British and French Channel Tunnel Companies were formed and in 1882 pilot tunnels were driven for nearly a mile under the sea in the dry Lower Chalk using the Beaumont tunnelling machine propelled by compressed air. The British Government then stopped the work for fear of a French invasion — not a good example of collaboration. The machine then presumably went to the Mersey.

The history of site investigation is affected by this work. De Gamond had tried to sample the sea bed by personal free dives of fifty seconds with little success. Among the many engineers who have been involved in this long running epic was Sir John Hawkshaw. He had borings sunk in the cliff at Saint Margaret's Bay in 1865. It is less well known that he entrusted a geological survey at sea to his young assistant Henry Marc Brunel (1842–1903) the second son of Isambard (Donovan D.T. 1966).

Young Brunel who was then 24 years old designed a gravity corer or drop-sampler consisting of a tube fixed below a stream-lined weight of 80 lb. This has been copied in all subsequent surveys. He carried out his work from a paddle steam tug, chiefly at night, in November and December 1866. He suffered the usual delays with the weather but he dropped his corer from 207 stations, all fixed by sextant to marks ashore. Thirty drops gave no result, 56 only superficial deposits, 10 proved Gault Clay and 111 proved Chalk.

Then in 1875–76 French engineers made a greatly extended survey at the request of the Channel Tunnel Companies. They obtained many thousand samples which were eventually stored in the waiting room of a Paris suburban station and were available in 1959.

b) The Start of Compressed Air Tunnelling

The First Hudson River Tunnel Part One (1878–1882)

The three sub-aqueous tunnels described above were all in good, fairly hard ground and *in the dry*— up to a point. The next is a very different story and full of admonition.

The idea of joining New Jersey to New York by railway tunnel had been considered for some years before D.C. Haskin made a start in 1879. This was the first serious attempt to drive a tunnel under compressed air and, in this case, it was attempted without a shield as this was nearly ten years before shield driving, as we know it, had come into practice. A.M. Hersent is credited with driving a small cast iron lined heading only 5 feet high and 4 feet wide with compressed air at Antwerp only a few months before, without a shield.

Between Cochrane's patent of 1830 and 1879 compressed air had been used to sink vertical caissons for bridges at Rochester in 1851 and at Saltash by I.K. Brunel in 1854, among others. So the use of air locks and the effect of air on different types of soil were known.

Haskin had a site investigation made for the Hudson Tunnel which revealed a considerable thickness of silt with some clay content which he hoped would embrace the whole of the tunnel. He decided that compressed air could hold up the silt and give ample time to build the brick linings for twin elliptical tunnels, 18 feet high and 16 feet wide.

In 1879 a 30 feet diameter brick lined shaft was sunk on the New Jersey side for a depth of 60 feet and the bottom sealed with concrete. Water in a stand pipe through the concrete rose to 9 feet below high water mark and rose and fell with the tide but only one third of the tidal range. (This agrees with other experiences recorded in a paper to the Second International Conference on Soil Mechanics, (Harding 1948)).

There is a long description of this work in the second edition of Drinker, written just before the project was first abandoned. So many lessons were learned the hard way that it can be quoted with advantage.

An opening was made in the shaft at only 23 feet below high tide level although the crown of the tunnel was to be much lower. A boiler air lock 6 feet in diameter and 15 feet long was built into the wall. At first the silt was firm and comparatively dry but after 3 or 4 days exposure it began to *run*. This was checked by putting on air pressure, at the end of 1879.

For the first four days digging actually went on with no protection whatever except the compressed air, so, naturally, blow holes began to form and the silt to run. On January 5th 1880 work started again with the use of segmental rings made of wrought iron plates, 2 feet 6 inches wide and 4 feet long with flanges made of 3 inch angles riveted to all four sides.

As the bottom of the air lock was above the crown of the tunnel a temporary sloping approach was made to reach the *break-up* for the North tunnel. A series of rings of iron plates, each being 18 inches bigger in diameter than the previous one were built until the full diameter of the tunnel was formed. The rings were joined at the top and the lower slope filled with concrete. For the running tunnel each ring consisted of 14 iron plates. In order to give space for the brick lining to form 18 feet internal diameter, the top three plates were 3 feet long and weighed 170 lb each while the rest were 6 feet long and each weighed 320 lb.

The intention had been to build one 2 feet 6 inch ring at a time and build the brick lining at once. But the silt stood up so well that optimism took over and, in a slap-happy way, the iron rings were advanced as much as 55 feet ahead of the brick lining.

"The face of the heading was always the exposed silt, which was so stiff under the air pressure that it could be cut in benches as a series of garden terraces and also into slopes rising from one terrace to another. An average slope of 45 degrees was usually left on this face and the excavation for and building of the rings always commenced at the top of the tunnel; usually 5 rings were building at the same time ... The bracing and timber ordinarily used in tunnelling were not required". (!) (Drinker 1878) **Figure 12**.

A novelty was the blowing-out of half of the silt into the shaft through a snorer pipe of six inch diameter after mixing 4 parts of silt to 1 of water. The other half of the silt was back-filled into the finished invert. The air pressure was said to be 18 lb per square inch with no mention of changes to suit the tide. No stand-by plant is mentioned.

Every twenty-four hours 83,000 cubic feet of free air was used from two compressors, a Clayton and an Ingersoll. "This amounted to about 125 cubic feet of normal air per hour for each man at work in the tunnel ... The air was quite pure and fresh, especially since electric light was introduced and the temperature was pretty uniformly 72 degrees". (Drinker) This is astonishing and may help to explain why Sir Ernest Moir wrote that the men working on this tunnel were dying at the rate of 25 per cent per annum, although 18 lb is not a high pressure. The present British Regulations call for 600 cubic feet per man per hour.

Work in the North tunnel went fast but carelessly, getting badly out of line and level. Drops in pressure caused blows which seem to have been sealed by throwing

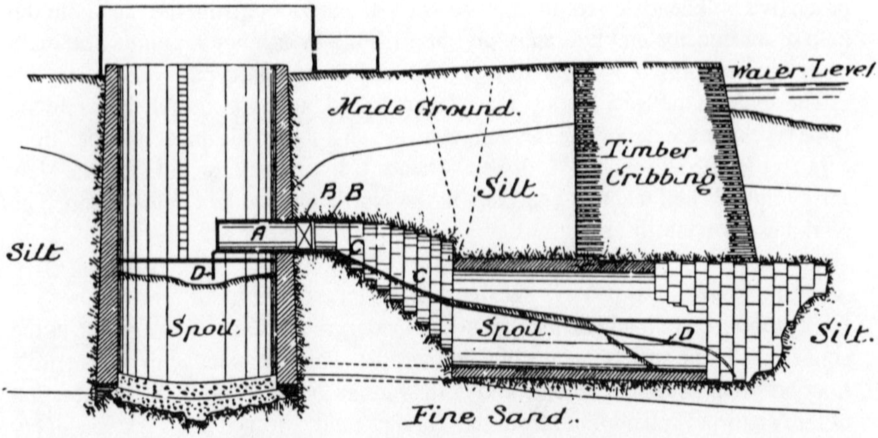

FIGURE 12. *Haskins first Hudson River Tunnel — Section (Copperthwaite)*

up silt, to be made firm by the pressure. It was fortunate that the face remained in silt and not sand and gravel. Then an attempt was made to build a connecting chamber against the shaft to embrace both tunnels. "The success attending the building of the first four rings induced the Superintendent to believe that it would be better to get in all the iron work in the chamber before bricking in the finished rings" ... The roof plates then became loose. "At 4:30 a.m. on July 21st the men left this plate and went into the lock. The Foreman, Burns, remained behind notwithstanding the fact that the men kept calling for him. The leak apparently broke out at the plate ... the men in the airlock came out and tried to stop the leak". There was a collapse and twenty men were drowned. Thus lessons are learned at a heavy price.

The work was recovered by sinking a caisson made of heavy timber covered with sheet lead and asphalt and equipped with an air lock for men and one for materials. The inner surface was a semi-circular arch to cover both tunnels and the caisson was lowered on screw rods. Under this protection the connecting chamber was completed round the wreck of the original entrance passage. Ingeniously, silt was blown from the caisson into the flooded tunnel. The compressed air stiffened it so tunnelling could be carried out inside the original tunnel in order to reach the distorted plates and connect the brick lining to the caisson work.

The Superintendent, Anderson, then took out a patent for the use of an iron lined pilot tunnel, 6 feet in diameter. This was about 35 feet long and driven 10 feet ahead while the back ring was rebuilt in the face as the main tunnel followed. It is reassuring to read that "every care was exercised in always protecting the face against any movement of the ground by temporary bulkheads well braced". In this way progress reached 6 feet per day. Air locks were built in the tunnels with protective bulkheads 4 feet thick. The silt was blown out from the face with the help of an injector and hydraulic pressure from a Worthington pump. The main rings were supported by radial timber props springing off the pilot lining.

The shaft on the New York side was then started but more sensibly by using the same type of arched roof caisson and, from the chamber made under its protection, a 74 feet length of tunnel was driven. Then in 1882 money ran out after the New Jersey tunnels had reached 1540 feet in the North and 600 feet in the South. The work lay dormant till 1889.

UNDERGROUND RAILWAYS AND THE FRESH PHASE OF SHIELD TUNNELLING

In 1853 a start was made on the first urban Underground Railway in London as the congestion in the streets had become intolerable. This coincided with Bazalgette's London Main Drainage scheme and both were accommodated in the reclamation of the Victoria Embankment from the Thames mud. Most of the work was in *cut and cover* under busy streets although several lengths were in tunnel. It was lined with brickwork and sometimes roofed with wrought iron girders, some of which have had to be replaced in recent years.

When completed, the Inner Circle connected main line terminii and became the prototype of many future *subways* or *Metros*. But the work created such havoc in the streets that a well-known civil engineer, P.W. Barlow visualised "a system of omnibus subways to be tunnelled at deep level so that the streets would not be disturbed. These may be small in diameter and worked by small cars pulled by cables" (Drinker 1878). **Figures 13 and 14**.

Sufficient wells had been sunk in London in the search for water to prove the considerable thickness of London Clay which underlay most of the city. This may have inspired Barlow who, in 1862 was sinking cast iron cylinders for the foundations of Lambeth Bridge in London Clay and this gave him the idea of a shield to advance in one piece, rather than the movable chambers in Brunel's invention.

In 1864 Barlow patented a circular iron shield to be used with a permanent lining of cast iron segments. He added "the space as it is left between the earth and the exterior of the tunnel may be filled by injecting, or running in fluid cement".

Barlow later took out a provisional patent for a much smaller, simpler shield strengthened by a series of diaphragms which left an octagonal opening in the centre for the workmen but capable of being closed at will to exclude water. Drinker and Sandstrom show this in error as used on the Tower Subway.

The Tower Subway (1869)

Parliament granted powers for a subway to be built under the River Thames at the Tower of London. James Henry Greathead who was an assistant to Barlow undertook this as a contract and designed a simpler shield propelled by screw

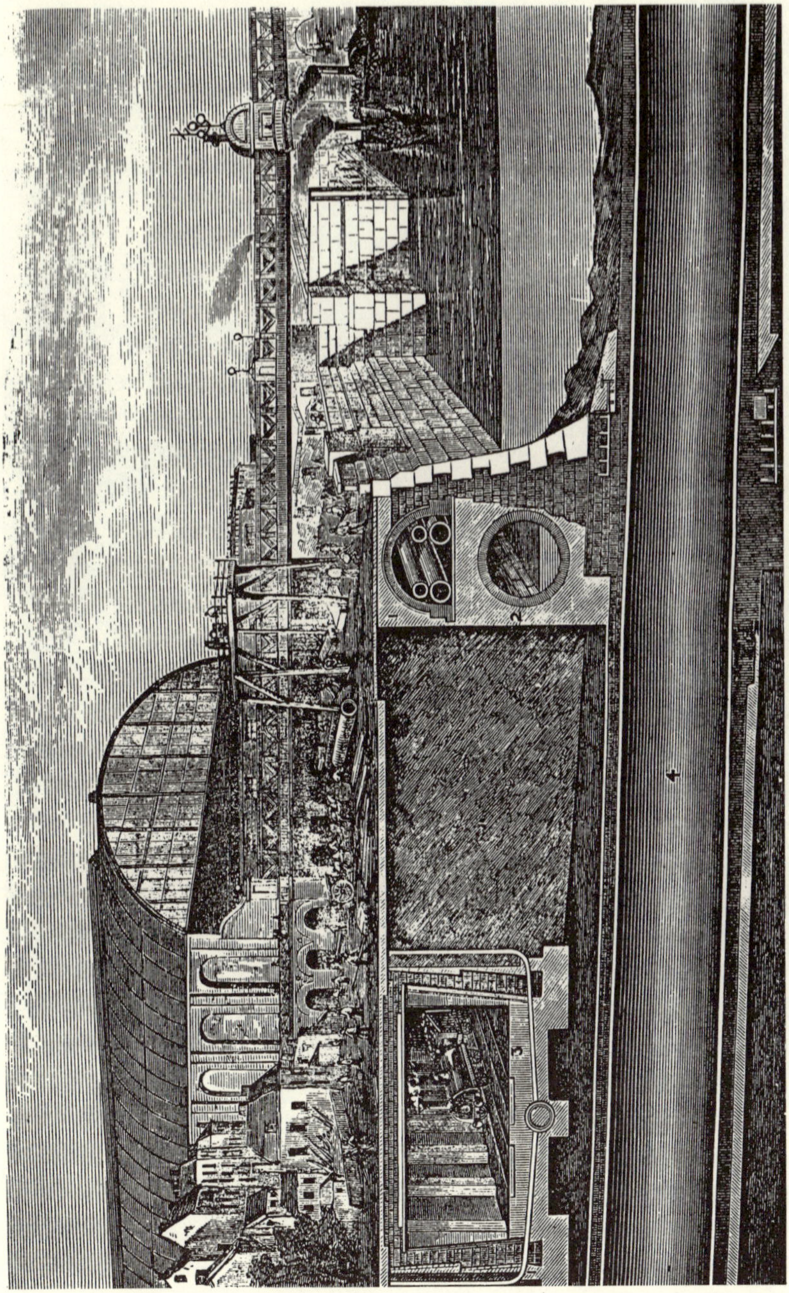

FIGURE 13. *Victoria Embankment showing low-level sewer and District Railway (this length of pneumatic tube started and abandoned at south bank).* (OLD & NEW LONDON)

FIGURE 14. *London Inner Circle railway cut-and-cover. (London Transport).*

jacks. This tunnel was driven in 1869 all the way in stiff London Clay so that the tunnelling itself was extremely simple but it set a pattern for the future, with the first actual use of cast iron lining, as suggested by Brunel.

The tunnel was about 1340 feet long. The lining consisted of 18 inch long rings of cast iron made up of three segments and a tapered key. The lining was only 6 feet 7¾ inches diameter inside and 7 feet 1¾ inches outside, though why is a mystery. The 9 feet diameter shafts were 57 feet deep and there was about 30 feet of cover between the top of the tunnel and the river bed.

Cement was injected behind the lining with a syringe. Copperthwaite (1906) says (p 9) that this was not satisfactory. During an air raid in December 1940 a bomb dropped above the tunnel. It only made a crater in the river bed of 7 feet in depth but the lining was broken and as the London Clay swelled it caused the segments to fold together round the water mains which were by then the occupants of the subway. This was repaired by driving 100 rings of 10 feet diameter cast iron lining round this length. The original metal of the segments was in excellent condition and the grouting behind the segments seemed quite successful (Harding 1945).

At the same time in the U.S.A., A.E. Beach built a shield inspired by Barlow's patent. It was 8 feet in diameter with a cast iron cutting edge, a central frame made of a heavy timber ring followed by a wrought iron forged ring to take the thrust of the hydraulic jacks, which worked under the protection of a steel tail which overlapped brick lining. It was used to drive a tunnel under Broadway to be used with cars propelled by compressed air in 1869 and is the first use of hydraulic jacks in a shield.

Shield tunnelling then went into cold storage until the start of the City and South London Railway in 1886. But much thought and pressures for more work continued. A number of patents were taken out, four by Greathead which included the grouting pan, of universal use ever since.

In 1876 he constructed a shield for a tunnel under the Thames at Woolwich which embodied a trap or water seal in between diaphragms and also a hydraulic segment erector. The work was not carried out but his ideas were followed in Britain and the U.S.A.

After Greathead's miniature triumph with his Tower Subway in 1869, this curious gap of seventeen years elapsed until the City and South London Railway set alight the remarkable outburst of shield tunnels, large and small, in the dry or underwater. A Beach type of shield was built in 1870 for use in a tunnel to draw water from Lake Erie to supply Cleveland, Ohio. This shield was 6 feet 6 inches inside diameter and 6 feet long. It was made of heavy boiler iron around two cast iron rings 4 inches thick and propelled by hydraulic jacks. But after 140 feet driven in very soft clay it became cracked and distorted and was abandoned.

FIGURE 15. *City & South London Railway — First electric locomotive. (London Transport)*

THE LONDON DEEP-LEVEL "TUBE" RAILWAY SYSTEM

In 1886 Greathead was able to start on the City & South London Railway after waiting sixteen years for a client. This twin deep level 10 feet 2 inch inside diameter, cast-iron lined, shield driven tunnel caused the real development in shield tunnelling and led to the big under-river shields described later. It was made possible by the sudden arrival of electric traction. Compressed air was used in some lengths in conjunction with a shield for the first time. This led to Pearson's Hudson Tunnel. **Figure 15**.

In London between 1886 and 1907 the basic network of seven rival lines, the result of private enterprise, had been completed and became united into the London Electric Railway Company. The diameter had been standardised at 11 feet 8¼ inches. One theory is that it was due to thinning the flanges from 11 feet 6 inches. Station shields were a simpler form of the large under-river prototypes. Attempts were made to use mechanical means of excavation, such as a Thompson ladder excavator working through the shield, but unsuccessfully. Then the Price rotary digger made its debut in 1897. This was simply a shield carrying rotary arms rotated by electric motors and carrying chisel picks. The average progress rose to 11 rings, 1 foot 8 inches long, in a twelve hour shift, compared with the usual four rings by shield. The contractor had to rebuild any ring more than one inch out of line at his own expense. As the Price Digger was difficult to control on piece-work driving some contractors found that it was not always economic for that reason.

The London Clay which is so hospitable for this form of work has a heavily eroded surface, often with deep depressions below three terraces, covered with water-bearing Thames sands and gravel. For this reason, and to get below foundations, the London Tubes are 70 to 100 feet below street level in most places.

Below the London Clay of varied thickness, lie the Reading Beds — also called Woolwich, because they outcrop around this London Borough. Tunnelling in them is costly so that the Tubes have not extended into South East London. Compressed air has to be used in the varied and often water-bearing beds below the London Clay as well as in the river gravels above.

The Greathead shield moves as one piece, unlike the ingenious devices in Brunel's shields, so the *assisted shield* method of working in the Thames sands and gravel became standardised. This consisted in laboriously scratching out a closely timbered chamber ahead into which the shield could be shoved. This needed all the precautions described earlier even with the help of compressed air.

Then about 1896 Sir Harley H. Dalrymple-Hay took out Patent No 622 for a hooded shield to be worked in conjunction with the use of clay filling in front. (Copperthwaite p 383). He was Resident Engineer at the time on the Waterloo and City tunnel (1894–1898) which was driven diagonally across and partly along the River Thames to carry business men directly from Waterloo Station to the Bank of England, the centre of the City. The tunnel is shallow for most if its length so there was much driving in the sands and gravel.

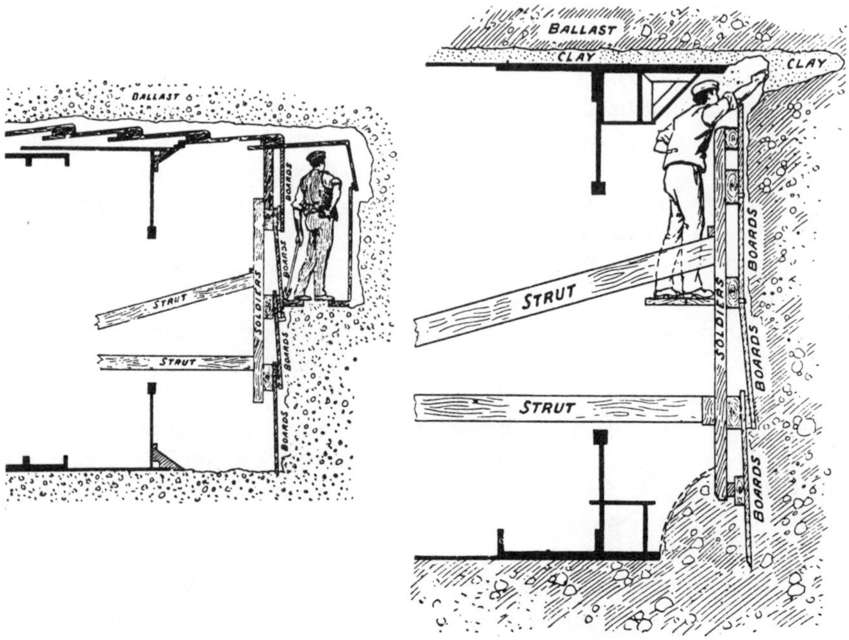

FIGURE 16. *Assisted shield and clay-pocketing method of driving. (Eng. Wonders)*

Dalrymple-Hay overcame the prejudice of the miners by a personal demonstration. A steel hood was built to extend the cutting edge of the shield by 2 feet, reaching down to *knee level*. The face was horizontally breasted, leaving a gap around the circumference of the cutting edge. Hay and another man started at the crown to scratch out holes about 2 feet ahead and quickly punned them full of puddled clay. Each then worked down either side making a series of clay pockets. **Figure 16**. The shield was then easily shoved with the hood entering the ring of clay pockets while the face boards remained static by strutting from the completed lining. It was then simple to advance the face, board by board, under the protection of the hood. The clay made a seal behind the lining, the consumption of compressed air was greatly reduced and the work speeded up. This revived Brunel's principle by using the whole shield as a movable overhead cover. In one case the air pressure held a miner's arm so tightly in the pocket that a miniature heading had to be driven round the arm to release it.

The shafts for this and the later *Bakerloo* line were sunk from timber piled stages in the River Thames. These provided working sites where materials could be brought in and spoil sent away by barge so that City streets were not cluttered with horse cartage. Future cooling water shafts and tunnels followed this lead. **Figure 17**.

FIGURE 17. *Waterloo & City Railway — Working site in frozen River Thames with twin shafts and air compressor shed. 1896 (John Mowlem & Co.)*

THE FIRST BIG UNDER-RIVER SHIELDS (1888–1908)

The first large sized shields since Brunel's for driving under rivers were triggered off by the success of the smaller Greathead shields used with such success on the City & South London Railway in 1886. It was also the first time that compressed air was used in conjunction with a shield on that line. The size and importance of the big shields allows them to take precedence over the story of the London *Tubes*.

It is also a curious fact that in London large diameter shields for the station tunnels came later, used on succeeding lines. The original stations were 35 feet in diameter to provide island platforms for the twin running tunnels. They were driven in the London Clay by hand and timbered on the English System, then lined with brickwork. The face was so large that four successive timber cills had to be underpinned in succession as the face worked down from the crown bars.

Figure 18 shows the sequence of the more notable shield drives of the period and how their number accelerated. The Sarnia, Hudson (Part 2) and East River Gas are the prototypes which pioneered this method by which all the subsequent tunnels were driven.

The Sarnia Tunnel (1888–1890)

The credit for the first big shield goes to the 2,000 yard tunnel under the St Clair river which joined the U.S.A. to Canada by rail. Two shields 21 feet in outside diameter were built with mild steel, now an accepted material. They were propelled by hydraulic jacks from a heavy segmental cast iron lining, something of a novelty. This shield, which owed more to Beach than Greathead was strengthened by a steel bulkhead which carried a mechanical segment-erector worked by hand. There were two bottom openings in the bulkhead, 6 feet high and 4 feet 6 inches wide which could be quickly closed by *guillotine* doors. This would give little comfort to men on the top stage. Fortunately the whole tunnel was in soft clay but a pressure of 28 lb per square inch was used to balance the potential head of 78 feet from invert to water level.

Two air compressors delivered into one six inch pipe, so it is not surprising that many cases of caisson disease occurred. Copperthwaite reports "At the commencement the miners were in the habit of locking out by means of a valve of 4 inches diameter but this was found too rapid and a 1½ inch one was substituted". One cause of carelessness is due to the fact that it is not always easy to lock in, but locking out causes almost no discomfort, yet rapid decompression is the root of the trouble. One of the lessons learnt over the years is the need for strict rules to enforce safety precautions, otherwise grown men have the childish urge to show *toughness*.

The First Hudson River Tunnel — Part Two (1889–1891)

In 1889 when fresh cash was available the British Contractor, S. Pearson & Sons

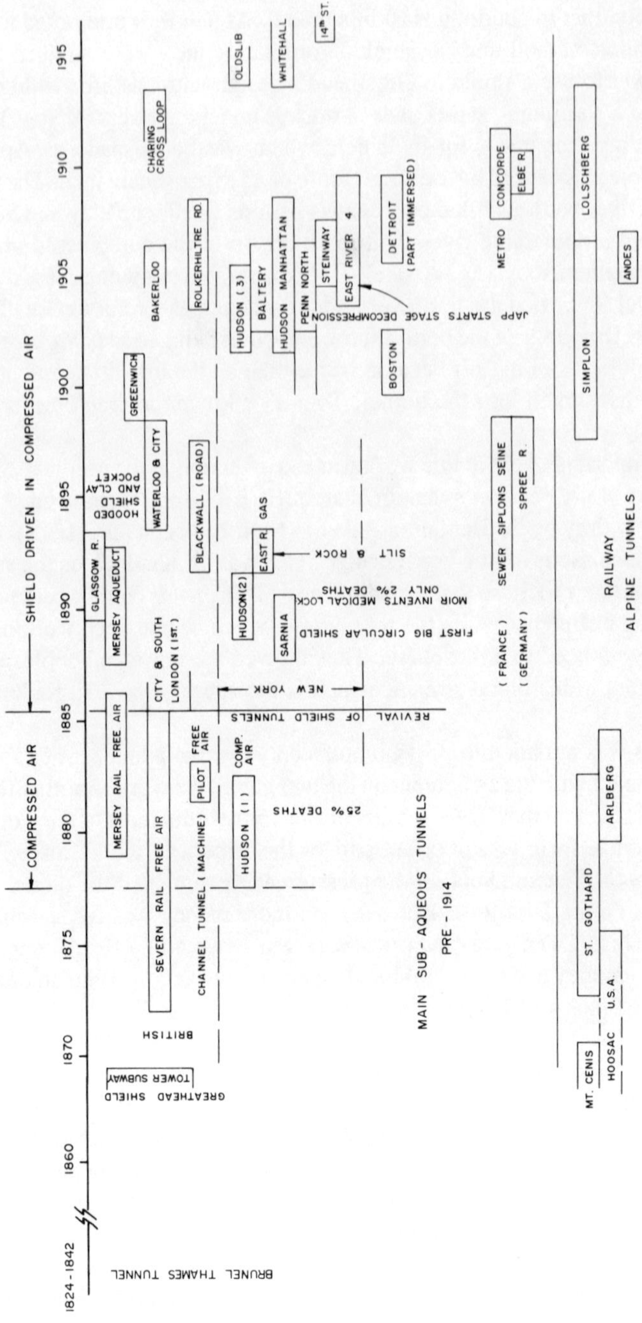

FIGURE 18. *Chart showing sequence of notable sub-aqueous and Alpine tunnels pre-1914.*

obtained the contract to continue Haskin's tunnel. At first they attempted to carry on with the pilot method and no shield. Fortunately they were advised by Sir Benjamin Baker to use a shield to Greathead's design with cast iron lining. This was the first of a continuing series of such tunnels under the rivers of New York.

While building a chamber for the shield which was being made by Arrols in Scotland, a *blow* was caused by the air pressure of 42 lb per square inch. The cavity formed in the river bed was filled by a canvas balloon filled with clay and hay. To erect the shield innumerable rivets had to be driven in the compressed air. The shield was strengthened by a heavy steel safety diaphragm but with access doors at each stage level. It carried the first-ever hydraulic segment erector to Greathead's patent of 1875. Drawings of the period show pit ponies being used *in the air* to haul spoil. The conditions of the silt became worse than on the first drive and at 35 lb pressure still discharged into the tunnel. Then in 1891 money ran out, as is the custom.

The most important event in this work was the invention of the Medical Airlock by (Sir) Ernest Moir, Pearson's man in charge. He had been recommended to the future Lord Cowdray by Sir Benjamin Baker as Moir had been his assistant on the compressed air caissons for the Forth Bridge. This was a boiler lock on the surface with its own entrance lock, so that a man suffering from *bends* could be attended to while he lay on a stretcher while the pressure in which he had been working was very gradually reduced to atmosphere. This allowed the nitrogen bubbles which had been formed in his blood stream, which is the cause of *bends*, gradually to dissolve.

The success was remarkable. In 15 months only two men died out of 120, or 2 percent, compared with the 25 percent on the first part of the work. Another feature of this tunnel is the fact that British engineers and miners did most of the work and many stayed on to form an important part of the American tunnel force. For a number of months Pearson kept on the pressure and maintained the tunnel at his own expense (Young D. 1966). But when no more money was forthcoming he boxed up the shield. Ten years later the work was resumed by British engineers with American money and the same shield again put into action. Pearson transferred Moir to the Blackwall Tunnel.

The Blackwall Tunnel (1892–1897)

This is the first sub-aqueous tunnel actually to carry road traffic, (though only designed for horses), in the heart of the London Docks. The cast iron lining, 25 feet inside diameter, 12 inches deep, was caulked with the old-fashioned rust and Sal-ammoniac mixture as each ring was built in order to save loss of air. The under-river length in compressed air is 3,116 feet.

Sir Ernest Moir used his experience on the Forth Bridge to sink two steel shafts 58 feet in diameter on the North Bank of the Thames, as compressed air caissons, using the air locks from the Forth Bridge. Two shafts were sunk on the South Bank by open grabbing. In order to save the time and cost of building the shield at the bottom of the shaft it was built on the surface in a dry dock during the sinking. The faces of the shield were blanked off with 4 inch timbers, the dry dock was flooded and the shield, which weighed 200 tons floated with 17 feet draught over the shaft. It was then lowered by pumping out the water. All four shafts had circular eyes closed by removable steel plates to allow the shield to pass.

The tunnel was designed to have only 5 feet of cover under the river bed. This reduced the pressure head at invert level to 80 feet and kept the restricted approaches to a grade suitable for horse traffic. The shield was driven in a straight line between each shaft and slewed to the line of the next drive. This was in order to save the cost of casting tapered rings of lining which otherwise would be needed on curves. Thus there is a kink at each shaft where the roadway is only 16 feet wide, which became bitterly regretted in later times as two commercial vehicles could not pass each other.

After the shield was landed the air deck was replaced in the shaft, using the vertical locks for access and the drive started. When sufficient length gave room a steel bulkhead was built which incorporated twin air locks at floor level and an emergency lock near the crown. The shield was unusually robust with twin bulkheads supported by cross beams providing four working platforms. Two air locks were built in the shield to serve the top platforms and two near the invert in the unrealised hope of using different pressures. Twelve chutes were also fixed at different levels for spoil to be shot out and they could be closed in emergency. The strata were very varied. For much of the drive the lower part was in the Woolwich and Reading beds in their customary variety.

For the last half under the river near the North bank a full face of sand and gravel was met in a buried channel. Once again a clay blanket was dumped, 450 feet long, 150 feet wide and 15 feet thick. This was encountered in the top of the face during careful and difficult work.

Moir again followed Brunel's lead by designing the shield so that the face, down to the bottom platform, could be kept completely closed by horizontal sliding steel shutters, 18 inches deep and secured by screw jacks. The screws were gradually released as the shield advanced so that the shutters held the face stationery while the shield was jacked forward. The sand and gravel was scraped or dug through gaps, made by moving the shutters or, in very bad ground, through smaller openings provided.

The shield carried two hydraulic segment erectors and, as another novelty, towed a timber stage with three floors for grouting, caulking and handling muck. Six compressors provided at the worst period 10,000 cubic feet per minute, which is suprisingly low for such an area of face in *Thames Ballast*. Two blows occurred which caused some temporary flooding in the invert. A heavy timber safety diaphragm was built in the top half of the tunnel with an emergency air lock near the crown and a timber walkway on one side at axis level. This was to provide an air pocket in case of loss of air. The heavy fog which immediately descends on sudden loss of pressure caused trouble when flooding occurred and men had to make their get-away. The air pressure was in general at 35 lb per square inch.

This tunnel provided many lessons, much experience and some novelties. The engineers were Binnie, Sir Benjamin Baker and Greathead. The Client was the London County Council who had a social conscience in advance of their time. They insisted that every man should be examined by their Resident Medical Officer who also took statistics which showed that the greater the quantity of air the less the cases of bends. Pearson's provided drying rooms and hot coffee. There were no deaths and few serious cases of illness.

East River Gas Tunnel (1892–1894)

This small 11 feet diameter tunnel becomes important as it was the first case of a mixed face of soft mud above and rock below. The experience gained soon became valuable in the New York tunnels which followed. The tunnel was expected to be in gneiss all the way but, as often happens, this at times changed into decomposed felspar in the form of a clay which decomposed in water. So compressed air was put on. The pressure was 48 lb per square inch yet water through fissures still liquefied the face and the contractor abandoned the struggle (Copperthwaite 1906).

The Engineer, an Englishman, Charles M. Jacobs then carried on by direct labour. On starting from the other New York shaft the ground proved to be in river mud, with rock in the lower half of the face. A shield was built and the rock was blasted in compressed air, while attempts were made to hold the mud above. At 45 lb per square inch pressure, connection kept being made with the river bed and live fish, crabs, mussels, boots, bricks and tin-cans entered the tunnel. The tunnel later passed into a face of rock but the shield was kept in use in case of more soft felspar, which did occur.

A Burst of New York Tunnels (1902–1909)

Then followed a pause of eight years until 1902 when, in another burst, six major under-river tunnels started within three years in New York, and others overseas (Hewett & Johannessen 1922). The lead was taken by the third attempt to complete Haskin's Hudson River Tunnel, in 1902. C.M. Jacobs as Engineer for the Hudson & Manhattan Railroad not only finished the North and South tunnels left by Pearson but carried on the line with 17 shields in bad ground under the foundations of New York City.

In 1903 Jacobs also became engineer for two more tunnels under the Hudson

River for the Pennsylvania Railroad, completed in 1906. In 1903 another tunnel for the Subway was driven under the East River which was completed by 1906 and called the Battery Tunnel.

Then in 1904 Pearson's were given the contract to drive four tunnels under the East River for the Pennsylvania Railroad which were finished in 1909. In 1905 the Steinway or Belmont tunnels were driven under the East River, close to the Gas Tunnel.

These tunnels shared the same problems. The faces were often in silt which varied in each river. In most of them mixed-face work was met in the face with soft ground above and rock below. Sands, gravels and sometimes a full face of rock added variety. Pressures varied from 20 to 40 lb per square inch. The American shields designed by Jacobs and others had a single diaphragm or bulkhead with a single segment erector and varying types of door.

On Pearson's East River tunnels the shields were almost identical with the heavy Blackwall shield with double bulkheads, air locks, muck chutes and sliding shutters, but as the sands and gravels proved less difficult than the Thames variety these extras were gradually dispensed with, including the sliding face shutters. A double-deck stage was towed behind the shield.

On Pearson's first Hudson drive the shield was shoved while allowing all the silt to enter, which then had to be dug away. Jacobs, on the South Hudson tunnel found that the shield could be shoved with the face closed and all the soft silt pushed aside, but this not only made fast progress but made the shield erratic in its drive. It was then found that if a certain amount of silt was allowed to enter, the line could be controlled by the proportion of silt admitted. This was adopted on the subsequent tunnels. The almost liquid silt became a stiff sausage under the air pressure as it squeezed through circular openings.

In the mixed faces the same methods were used as in the Gas Tunnel. Usually the ground was dug and timbered ahead of the hood of the shield and the shield was shoved into the excavated area while the breasted face was kept in position by strutting from the tunnel lining or by exhausting hydraulic jacks as the shove went on.

In most of these tunnels many tons of clay were dumped ahead of the drive usually to forestall trouble. In Jacob's Hudson tunnel it is recorded that some of the dumped clay which entered the top of one shield would not respond to the air so it was baked hard by blow pipes fed from tanks of kerosene, installed at the face. As fire is a dangerous hazard in compressed air this is an expedient that is best forgotten.

There were fears that tunnels would settle in such soft material and an attempt was made to anchor and support them by piles driven through openings made by removing bottom plates. Screw piles in the Penn Hudson and jetted tube piles in the

Battery were tried but it was found that the tubes floated safely in the silt without such aids. The work was necessarily hazardous and legends have arisen, but it seems true that one man was blown through the face into the river and survived.

One big step forward was the work of Pearsons' engineer in charge of the East River tunnels, (Sir) Henry Japp, who introduced stage decompression by using air locks in series with a short walk between each. This resulted in no deaths and no severe cases of caisson disease.

After this spate of tunnels there was a pause and then they become almost commonplace, including the more recent large road tunnels in England and elsewhere, but these tunnels were the training ground for much that followed.

In those days soil sizing analyses and other tests were not available. Descriptions of the silt are scattered in several chapters of Hewett & Johannessen. (Hewett was later Resident Engineer on the first Mersey Road Tunnel). The silt under the Hudson is described as an extremely finely divided dust supersaturated with water. Its passive resistance is destroyed by the movement of the shield but it gradually returns to its original condition. The water content is given as 55 per cent by volume.

The Rotherhithe Tunnel (1904–1908)

One more tunnel deserves notice in this section, mainly because it proved simple, partly because lessons were learned but also because the ground, for once, was benevolent. The Rotherhithe Tunnel runs diagonally across the Thames only one quarter of a mile below Brunel's Thames Tunnel and in identical strata. In this case clay cover, although only 8 feet thick, remained above the crown of the tunnel for the under-river portion. This produced completely different working conditions from Blackwall. The strata in the lower part were again the Woolwich & Reading beds but the air pressure kept them under control. **Figure 19**.

The novelty was the fact that the contractors, Messrs Price & Reeve decided to use their Price Rotary Digger shield, which they had introduced on the Central Line Tube in 1897, to drive a pilot tunnel ahead of the shield at a rate of 13½ feet per day. The cast iron lining was recovered for future use. The shield was much simpler than the Blackwall shield but it had double bulkheads which were designed to provide water traps for each compartment. There were 16 of these, formed by three vertical and three horizontal diaphragms for the whole area of the shield. There were two hydraulic jacks in each compartment for holding the face boards during shoves.

A travelling stage was used of two levels. Progress averaged 5½ feet per day and pressures ranged between 14 and 20 lb per square inch. This shows how quite a small variation in ground can completely change driving conditions in different tunnels.

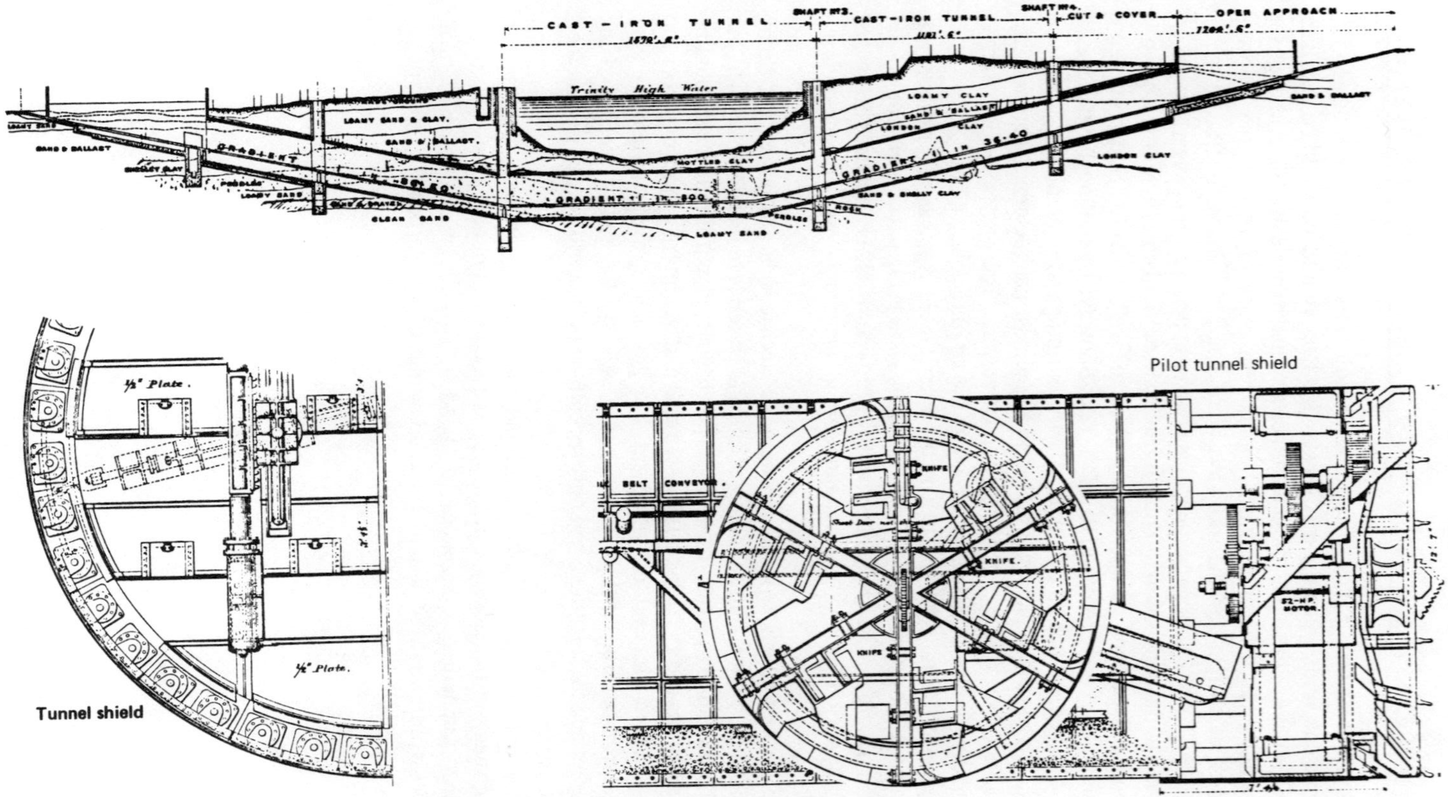

FIGURE 19. *Rotherhithe Tunnel — Section showing strata, Price tunnelling machine for pilot tunnel and main shield. (I.C.E. Brunel Exhibition)*

Both this and the Blackwall Tunnel were built in the days of horse transport but had spacious voids under the roadway so that later ventilation could be installed. But in the days of the horse, exhaust fumes also had to be dealt with. Although the Rotherhithe was 27 feet 8 inches diameter in place of 25 feet, the roadway remained at 16 feet and the footways were 4 feet 8½ inches instead of 3 feet 7½inches. These fractional dimensions keep re-appearing for no clear reason.

Owing to the congestion of horse traffic, other cities began to follow London's lead and underground railways were started in Glasgow, 1891, Budapest, 1893, Boston, 1895, Paris, 1898 and the New York Subway, 1900, though much was in cut and cover and not all by shield. Cast iron linings were dominant, with local differences in detail, but this also led to the almost complete elimination of the tunnel bricklayer whose craft was diverted to sewers. Aqueducts and comparatively small hydro-electric schemes carried on the art of rock tunnelling.

The period from 1886 to 1914 is the First Shield Age, as this method was dominant. Cities were growing rapidly, roads were being paved and people were becoming more civilised with the novelty of bathrooms and lavatories. This caused a fresh attack on sewers, drainage tunnels and water supply all over the civilised world.

The greatest advances in this epoch were electric light and power, providing locomotives, cranes, pumps, compressors and batteries to make life simpler, but also the invention of a new race — the Electrician. Plentiful mild steel could be handled by electric and oxy-acetylene burning and welding. Supplies of rubber and petrol emerged with eventual influences. The gas engine developed into the other internal combustion types of petrol and diesel. By the end of the period the typewriter and the telephone had only recently appeared and aircraft were in their infancy.

Then came the 1914–18 War which put a stop to work in many countries. In Britain many of the closely knit body of Engineers and miners joined the Royal Engineer Tunnelling Companies to spend a hard life undermining German trenches by primitive methods as did their German counterparts (Pequignot 1963).

Chapter 5

MODERN TIMES

THE TWO DECADES BETWEEN THE WARS: 1919–1939

At first slumps and labour troubles delayed fresh advances in Europe. Then in London there came a second burst of tube railway extensions. In the first decade actual techniques in clay changed little except in details. The introduction of compressed air tools such as hand-held spaders and jackhammers increased progress once prejudice was overcome. This led to the abandoning of the timbered box heading ahead of the shield with the saving of at least four men in the gang. Petrol driven trucks and lorries began to replace the horse drawn tipping carts which, until then was the only way to cart spoil through the streets.

At this time other aids appeared. Steel sheet piling gradually evolved which revolutionised work in cofferdams and shafts. About 1930 the first bored piles about 12 inches in diameter came into use. Self-priming and submersible pumps made life simpler especially in shallower tunnels such as the greatly extending systems of sewers in all parts.

The extensions brought new and interesting complex work. Lines became interconnected for the first time which needed flying junctions below ground. These were built each night in the four hours when the trains were not running. These are called step-plate junctions as existing running tunnels are widened out on one side by groups of eccentric rings, increasing in diameter from 12 feet up to 26 feet. This size allows a new 12 foot diameter running tunnel to be driven to enter the junction chamber beside the existing one. Some new stations were built, while the trains continued to run, by excavating and building 22 feet 4 inch rings of lining around the running tunnel. Each existing ring carrying the track had to be carefully supported on timber trestles as the work went forward. In one case a 27 feet 6 inch diameter cross-over tunnel was built using 15 lb per square inch pressure. The old grout holes had to be plugged so that escaping air would not embarrass passengers.

The lesson of the risk of tunnelling near the top of an unevenly eroded bed of clay was re-learned when widening the original City & South London railway tunnels between traffic hours. Although borings had been made through the old lining at intervals they failed to reveal a sudden depression filled with sand and gravel between two of them. At 5 p.m. a train entering a station was seen to be carrying a dislodged poling board. Traffic was stopped and gravel and water was seen to be trickling in. Half an hour later 500 cubic yards of sand and gravel had run into the tunnel causing a crater 45 feet in diameter and 12 feet deep under the roadway. At 6 p.m. a gas main in the crater exploded and blew up the roadway.

The crater was duly filled with sand and gravel and a number of one inch diameter perforated pipes driven into the fill which was then solidified by the Cementation process. Not surprisingly work then went on for 24 hours a day without trains (Jones & Curry 1927).

A curious work started during, and finished after, the war is the Post Office Railway, a cast iron lined tunnel driven from many shafts across London from Paddington to Liverpool Street, with loops and step plate junctions at each main Post Office, for handling parcels and mail. It is 6½ miles long, 9 feet in diameter and carries twin tracks of 2 feet gauge, with 25 feet diameter stations. It is the first fully automated system, with driverless locomotives.

In 1930 the disastrous slump hit many countries. To reduce its effect the London Electric railway embarked on a further programme of improving their stations, now that the escalator had come into its own. This called for complex and ingenious works, not only in driving large escalator tunnels at 30 degrees to the horizontal, sometimes uphill, and the landing chambers, but also underpinnings, sewer diversions, cable tunnels, cross passages, driving bridges in tunnel across existing station tunnels, as well as making new large booking halls. All this gave experience beyond that of continuous long tunnels.

This coincided with the introduction of new Geotechnical Processes notably of the Joosten two-chemical injection process and ground-water lowering by bored filter wells in Europe, and of Well Points and AM9 in the U.S.A. (Harding 1946/7). Chambers up to 26 feet in diameter had to be excavated by hand labour. Pilot tunnels were used to break up the face so that supporting was simpler. But at the end of a chamber the full face of London clay had to be held. Large timber *soldiers* had to be erected and moved forward to support face struts and walings. The small, newly invented compressed air tugger hoists were invaluable in helping to handle such bulky items in a confined space, using pulley blocks shackled to bolt holes in top segments.

Then in 1935 as Governments plucked up courage a third burst of *tube* extensions took place in London, of which the Central Line eastward involved most tunnelling. This included ten shields driven in compressed air under the Lea River and its many tributaries in the same strata as the Thames and Rotherhithe tunnels, but in this case every one of the new expedients was used. In 1938 rearmament led to a rocketing of the price of cast iron so the 40 year monopoly of cast iron linings was broken by the introduction of flanged reinforced-concrete segments, which have become a permanent feature in many works.

In other countries metro extension went on. In Berlin ground-water lowering by bored filter wells made it possible to work in their water-bearing sands. Much footage was made in the World's Press over the new Moscow Underground which started about 1932. The size and decor of the stations show that prestige was more important than economy.

In the discussion of the Ilford Tube (Groves 1946) Sir David Anderson described the visit of a British Mission of Underground Engineers to Moscow at the start of the work at the request of the Russian Government. They gave much free advice on tunnelling and running the system. Messrs Markham sold the Russians a single tunnel shield and erected it for them, after which it quickly sired a numerous Russian progeny.

Under-river road tunnels increased such as the Holland under the Hudson River, opened in 1927 and the first Lincoln in 1937, in compressed air. In England, the first Mersey road tunnel, the Queensway, added variety from its size — it is 45 feet internal diameter with four traffic lanes in a roadway of 36 feet. It was driven in red fissured sandstone (1925–1934) in free air for 3 miles in length. To cope with water, as cement injections had not been successful in closing fissures, a drainage tunnel in the form of a seven foot heading was driven from one bank to meet the downhill drive from the other. **Figures 20 and 21**.

Two headings 15 feet by 12 feet were driven, one close to the crown and the other in the invert. From these headings several break-ups were made. Part of the length was lined with heavy cast iron segments and part with circles of rolled steel joists, 2 feet 6 inches apart with concrete between them. A hanging stage from the finished linings kept the top heading in use for construction traffic, while spoil was removed through the bottom heading. The magnitude of the ventilation problem and the works to house the system for such a large tunnel was overcome by studying the experience of the Holland Tunnel.

In 1937 a start was made on a road tunnel under the River Thames at Dartford. A 12 feet diameter pilot tunnel was first driven using high air pressure in the fissured Upper Chalk. At one time the installed capacity of 8,000 cubic feet per minute was needed. Ten thousand tons of cement was eventually injected but details belong to the post-war story. Shields were built in chambers under each bank of the river but the war started in 1939 and the shields were *moth balled* and maintained until work resumed in 1957. The shields are almost a replica of those on the Rotherhithe tunnel in 1907.

Rock tunnels improved with the adoption of horse-shoe arches of steel joists set at suitable (or unsuitable) intervals. Explosives and methods of ventilation improved. Conveyors and mechanical loaders became more popular. Battery locos took the place of those powered from an overhead wire. Access and transport remained mostly by rail though the new-found ability to make a diesel motor consume its own smoke led to the possibility of running trucks on the tunnel floor. Detachable bits and better hollow drill steels made drilling easier.

Tunnels for hydro-electric power began to increase though these were usually

FIGURE 20. *First Mersey road tunnel (Queensway) 1931 — 200 ton semi-circular roof shield in sandstone rock under Dale Street, Liverpool. (Mott, Hay & Anderson)*

FIGURE 21. *Mersey (Queensway) Tunnel — Excavating invert to full circle under river in sandstone from upper and lower pilot tunnels. (Mott, Hay & Anderson)*

driven through mountains before delivering water down steel pipes on the hillside to power stations built in the open at the foot of the slope. In the U.S.A. among other works the vast tunnels involved in what was then the Boulder Dam, caught the public imagination. Mountainous countries with no fossil fuels like Sweden and Switzerland found that Hydro electric power was a necessity and began to lead the field. In Scotland the Lochaber scheme (1925–30) with 15 miles of tunnel provided power for smelting aluminium, soon to become a war-time need. The Galloway scheme in Scotland followed (in 1931–37). Large underground Power Stations come later. Dr. Jaeger tabulates only three of more than 75 MW before the Second World War (Pequignot 1963).

Four immersed tunnels belong to this period, all for road traffic. Three in North America, Posey, Detroit-Windsor and Bankhead were the customary two-track circular steel units. The fourth was the Maas at Rotterdam started in 1938. It took the form of reinforced concrete boxes in three compartments which provided a space for cyclists and cables and four lanes of road, two in each compartment. The units were joined after laying by compressed air caissons and a means of pumping sand beneath them was devised. Large compressed air concrete caissons were sunk on each bank. The caissons contained the ventilation plant and formed a transition between the open cut approaches and the tunnel. The work was started by a consortium of Danish, Dutch and German engineers. It was completed during the war under Germans by captive Dutch and Danish engineers. During the second decade (1930–39) the Gospel of Soil Mechanics preached by Dr Terzaghi at last began to penetrate the entrenched mind of the Civil Engineering Profession.

Looking back over this period there is a contrast between progress on the surface and below ground. In 1919 the horse was much in use, even for pulling scrapers in the U.S.A. The only excavator in general use was the steam face-shovel which dug the Panama Canal while steam and newer electric cranes ran on rails.

By 1939 caterpillar tracks, developed from tank warfare were in universal use on the whole gamut of tractor-propelled *muck shifting* plant as well as the dragline and all its variations.

This had little effect below ground, where, in Britain if not elsewhere conservatism had to be overcome as well as the reluctance to spend money at all, let alone wisely. Most winding was done from tunnel shafts by crane and in some circumstances by Scotch Derrick which had not been superceded by the continental tower cranes of the 1950's. Some firms began to use primitive tower hoists which were accident prone.

The advent of the *Geotechnical Process* began to produce a change. At first wishful thinkers hoped for a magic cure for all their troubles until they learned the lesson that all such aids are expensive to apply. Nature continued its tuition. If the ground is difficult it will prove expensive to tunnel, which must be paid for in money and time whatever expedient is used.

WORLD WAR TWO (1939–1945)

In 1938 when war seemed likely, countries who would suffer put defence works in hand, such as tunnels for storing oil and vital materials, shelters and command posts. In Britain tunnel engineers were allocated zones of responsibility for repairing underground services after air-raid damage. As a precaution heavy steel flood gates were installed in the Tube stations to prevent flooding if under-river tunnels were broken (Civil Eng. in War 1948).

As the war escalated much further work was put in hand by both sides. Underground factories in Germany specialised in slave labour. With some foresight, shelters under London were driven as 16 feet diameter tunnels lined with concrete segments, driven on transition curves so that they could later form part of express lines. But change of policy prevented this after the war.

In place of construction, engineers had to learn to repair destruction, which differed in different places. The massive retaliation raids on Germany must have posed different problems from those in London in the 1940–41 raids. These raids went on nightly for seven months, but it was just possible to keep vital services such as sewers, water mains, cables and trains from completely breaking down. Daily life could then continue — after a fashion — as the raids were wide spread causing separate *incidents*.

Deep cast iron lined tunnels suffered less than shallower brick and cast iron sewers but there were cases where the bomb shattered massive rings of cast iron by shock through the clay, without penetrating. In some cases the lining collapsed allowing blocks of solid clay to drop into a station. This allowed the waterbearing sands and gravel above to flood the tunnel, assisted by burst water mains and so burying shelterers. As bombs do not fall vertically the Bomb Disposal units of the Royal Engineers had to relearn the method of the classical timber heading when following the trace of unexploded bombs. In the worst case a heading had to be driven for over 70 feet to reach the bomb.

In six years of war, conflict does not go on every day, and mental activity increased in many countries. In Britain the Institution of Civil Engineers was very active as members near London had little else to divert them. Following the lead of the Building Research Station many papers were discussed on soil mechanics and tunnelling, with much full scale research.

Within a few days of the end of the war in Europe the Allies sent missions of experts to find what could be learned from the work of the Germans, including such novelties as electro-osmosis. Leading German experts were invited forcibly to take up residence in allied countries to pass on their expertise. Curiously this led to the U.S.A. reaching the Moon and also Russian rocketry. But as rockets go up, Man again goes below to escape their inevitable descent.

AFTER THE SECOND WAR: 1945–1976

Much has happened in these three decades compared with the simple but important advances of the 19th Century. A notable change has taken place in the mental processes of engineers, and — up to a point — of Directors and Bureaucrats.

The most distinctive advance has been in the science of Soil Mechanics, Rock Mechanics and the understanding of Geotechnical Processes, leading to a revolution in the methods of site investigation. In Britain this ceased to be the reluctant side-line of well-sinking firms of mediocrity. In 1945 the first British firm, Soil Mechanics Ltd., a subsidiary of John Mowlem & Co., was founded which had its own boring department and laboratory, staffed by qualified engineers with experience in actual construction. By 1970 over forty firms of differing types were competing for site investigation work. The foolishness of such competition has not yet reached the heads of Officials.

A dominant feature of this period was the growth of National and International Societies devoted to classes of work. Those for Soil Mechanics and Foundation Engineering and I.C.O.L.D. (International Conference on Large Dams) led the field. For the first time Organisations paid the expenses for members of their staffs to travel to such conferences, with fruitful results. This would have shocked pre-war employers to the very bottom of their pockets. A late addition to these was the International Tunnelling Society, founded in 1974 by many National Societies.

Engineering Geologists began to be employed on the staff of firms where none existed before. The experience accumulated in these thirty years is massive compared with the ignorances of predecessors. All these improvements have greatly affected the art of tunnelling. But newer knowledge should be tempered with the lessons of the past.

This became the age of plastics, resins and other materials based on carbohydrates, automation, computers and transistors. The fact that Man could fly anywhere in the world commercially had many side effects over progress. So new methods and improvements were sought after, which overlap into different types of tunnel and of ground.

These new methods include newer types of lining in concrete, welded steel and special iron, new plant for concrete placing, shotcrete lining, rock bolting, ground anchors, diaphragm walls, and bored piles of ever increasing diameter, loading devices, transporting methods, new uses for conveyors, tunnelling machines in rock, new ways of drilling and blasting, new types of hydraulic mechanisms with finger-tip controls with varieties of plant for all purposes.

Other advances were new materials for injection processes and methods such as tube-à-manchette, taking freezing into tunnels where for years it had been reserved for occasional shafts, better submersible and self priming pumps with new materials for tubes and pipes and their couplings, geophysical methods over water, better boring and sampling devices with more use of diamond drilling, more testing techniques in borings, stress measurements and uses of X-ray and closed circuit television.

In tunnelling itself there were many more hydro-electric schemes, new Metros

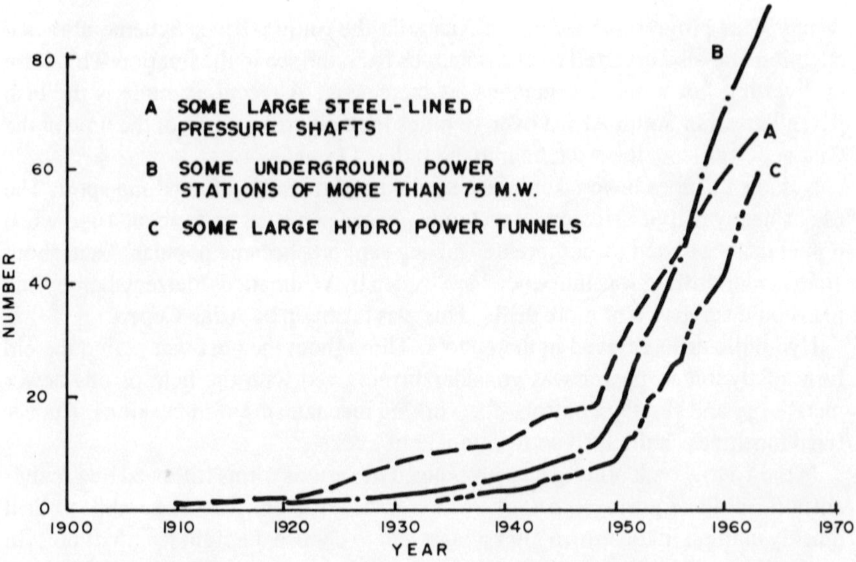

FIGURE 22. *Chart showing rate of increase in hydro-electric tunnel work. (from tables in Pequignot)*

or subways in many cities, immersed tubes, new techniques in deep shaft sinking arising from the Orange Free State Goldfields with strange new uses such as the mile-long circular tunnel at Geneva for the Europa Atom Splitting Research Project. With all this there is a proliferation of words, articles and books, Codes of Practice and Government interference.

One consequence of this expansion is to re-emphasise how much was achieved by our predecessors — by Horse and Man, pick and shovel, axe and saw and the Blacksmith's forge.

The increase in number and variety of works over these three decades has been kaleidoscopic.

Rock Tunnelling

The increase in rock tunnelling gave a boost to the Science of Rock Mechanics, which includes all previous but untitled knowledge and experience. The crescendo of Hydro-electric work all over the world is shown in **Figure 22** which summarises three tables compiled by Dr Charles Jaeger (Pequignot 1963). Before 1940 he lists only one hydro-power tunnel, three pressure shafts and three underground power-stations of more than 75 MW. By 1964 there are 50, 62 and 93 respectively. Since then many schemes, both vast and minor, have been embarked upon, notably in the U.S.A. and Canada, a newer variation is Pumped Storage.

In Britain the Scottish Hydro-electric board started work after the war on a twenty year programme, while in Australia the Snowy River Scheme attracted attention and also diverted river discharges from the sea to the interior. This form of diverting flows under watersheds is increasing. A recent example is the Fish River Tunnel in South Africa over 50 miles long diverting much of the flow of the Orange River from the Atlantic into the Indian Ocean.

Notable changes have taken place in plant and methods of handling spoil. The use of heavy drifter drills on large Jumbo stages persisted until about 1950 when lighter drills assisted by compressed air-leg supports became popular. Then about 1960 ladder drilling was introduced in Sweden by Widmark & Platzer whereby one man could serve two or more drills. This was taken up by Atlas-Copco.

Hydraulic drills revived in the 1970's. Throughout the post-war period the old form of hydraulic power was considerably revived with the help of oil, newer metallurgy and simpler controls, for working mechanisms of many kinds. Rubber tyred machines challenged conventional rail tracks.

In the 1950's rock bolting was introduced in various forms followed by ground-anchors, both requiring types of drilling. The problem was to be able to drill quickly enough in the crown after a blast and to choose the right length of bolt. In some underground Power-Houses bolts ten feet long had to be changed for thirty feet after roof collapses.

Arches of steel joists continued as a form of support though less conventional minds began to use precast segments to provide temporary and permanent support simultaneously.

Rock tunnelling in the Alps revived to provide road tunnels to relieve the historic passes after some road traffic had been diverted to the old famous railway tunnels by the use of special carrier trucks. The main ones are Mont Blanc, Great St Bernard, Brenner, Simplon and St Gothard. In spite of all modern conveniences the rocks themselves produced the same troubles as the railway tunnels, with intense rock pressures, decomposed and swelling rock, with water, both hot and cold.

The Mont Blanc tunnel, started in 1958 is a good example. The French adopted the American practice of going for a full face with heavy, railborne plant with heavy drills, and holes up to 8 inches in diameter. The Italians favoured lighter plant and a more piece-meal attack on the changing face. Both methods worked. On the French side massive protogine granite, a post-crystalline deformed rock that rapidly loses cohesion under rock pressure after firing, was dealt with by 465,000 yards of rock bolts. The work took seven years to complete.

As an example of the variety of ground, in the Great St Bernard tunnel, opened in 1964, micaceous schist, distorted almost vertically, was met on the Swiss side and a brittle carboniferous schist on the Italian. In a hydro tunnel high up above Mont Cenis a metamorphic sedimentary rock was found in horizontal layers, due to the folding over of 180 degrees.

As a variation in rock tunnel needs — the Dokan Dam in Iraq (1954–1958) required not only 12 metre diameter tunnels for river diversions and flood disposal but also tunnels for carrying out one of the largest grout curtains yet undertaken. These tunnels were 3.55 metres high and 2.45 metres wide, driven in the permeable dolomite. The cost of the tunnels was balanced against the saving by the ability to drill much shorter holes and by not hindering the construction work above. Seventy-seven thousand, seven hundred and sixty-six metric tonnes of cement and sand were injected through 601,000 feet of drilled holes. (Binnie et al 1959).

Tunnels Driven Under Water

These multiplied greatly to fill different needs. Many were in compressed air with the use of a shield while some were in rock. Owing to the steady demand for power ever larger Thermic Power Stations were built which needed many tunnels for cooling water and for carrying high-tension cables under rivers. In 1950 the advent of the Atomic Power Stations added to the number of tunnels which were driven under the sea. At Dungeness and Sizewell in Britain, intake shafts were jacked up through the sandy sea bed from inside the tunnel. Special steel segments were built as part of the tunnel lining which became caps to the shaft so that it could be jacked up through the opening which they formed. As the tunnels were in compressed air restraining anchors were needed in case the shaft tried to emulate a Polaris Missile. The introduction of jacked-up platforms evolved for the oil industry, aided much work at sea.

Sewer outfalls increased. In some cases in chalk or soft rock injections of cement and chemicals took the place of compressed air where high pressures could not be coped with.

Road tunnels under water multiplied. The first in New York, under the Hudson River was the Holland in 1927, followed by the three Lincoln tunnels opened in 1937, 1945 and 1957. They closely followed the pattern of the famous rail tunnels in much the same conditions.

In Britain the Government was at last converted to the use of tolls which led to a programme of tunnels and bridges. The Dartford tunnel shields were unwrapped in 1956 after 20 years below ground. Before the 1939 war ten thousand tons of cement had been injected through grout holes in the 12 feet diameter pilot tunnel assessed as 5 cwt per cubic yard treated, in the Upper Chalk. In 1956 injections of clay-cement and clay-chemicals were carried out from staging through the new tube-à-manchette technique to treat the gravel on each bank where the tunnel emerged from the chalk into the gravels. 305,000 cubic feet of various grouts were used to treat 30,000 cubic yards of ground. The average air consumption in the untreated 12 feet pilot tunnel was 4,000 cubic feet per minute or 30 cubic feet per square foot of face. For the 30 feet O.D. treated tunnel the average was 1,500 cubic feet per minute or 2.1 cubic feet per square foot of face. (Kell J. 1963).

The start of the Dartford tunnel was quickly followed by that of the twin Clyde

FIGURE 23. *Mersey (Kingsway) Tunnel, 1967 — Concrete lining with welded steel plates for watertightness. (Mott, Hay & Anderson)*

tunnels (1957–1964) shield-driven in compressed air through shales and sandstones, underlying glacial and river deposits. Injection of bentonite cement and chemicals simplified the work. Soft ground with rock in the lower half was met in the tunnel face. (Morgan H.D. et al 1977). The tunnel under the river Tyne (1961–1966) was constructed mostly in rock, as an earlier tunnel for pedestrians and cyclists showed this to be desirable.

Lessons were learned from the second Blackwall tunnel (1960–1965) using two of the hooded shields from the Dartford tunnel and with an even more elaborate system of injection work. This tunnel was deeper and further downstream than the 1892 tunnel so higher pressures were expected in worse ground. Firstly two pilot tunnels 7 feet in diameter were driven, using hooded shields. Rapid progress was made by using high pressures and much consumption of air without clay pocketing. This reached 9,000 cubic feet per minute or 180 cubic feet per square foot of tunnel face.

One pilot tunnel was in the crown and the other in the invert of the future tunnel. Gland valves were screwed into 3 inch holes provided in the 7 feet lining so that, after the pressure was taken off, injection pipes could be driven through them in free air. The whole area around the future tunnel was impregnated by "a proprietary low-viscosity chemical grout", i.e. T.D.M. The air consumption in the 28 feet 2 inch tunnel was only 5,500 cubic feet per minute or 8 cubic feet per square foot of tunnel face. The maximum pressure reached was 36 lb per square inch (Kell & Ridley 1966).

The Kingsway is the title of the twin tunnels under the River Mersey started in 1967 in the red sandstone of the area. They were driven by the Robbins machine which had driven 5 tunnels at the Mangla Dam after adaptation. One lesson was in the need to renew the main bearing of the Mole in such cramped conditions. A zone of weakness was revealed in one pilot tunnel which required rock bolting. As the main Mole had to pass later, bamboo rods were used in place of steel with success. Another innovation was the lining which was made of concrete blocks but cast with a steel plate as the inner face. These plates could be welded to make the tunnel waterproof. **Figure 23**. The invert segments were 12 feet long to provide a flat invert for travelling during construction. Then the duplication of the Dartford tunnel started in 1974. (Megaw T.M. et al 1972).

Tunnelling under the sea has been the habit of coal miners for many years where the carboniferous rocks above the coal seams are reasonably impervious. This greatly increased in Britain after the Coal Board had commissioned an ingenious drilling tower for boring at sea before the coming of the jack-up platforms.

Two other tunnels under the sea deserve comment. In 1957 the U.S. Navy released the secrets of geophysical work at sea in time for the big surveys carried out by the Channel Tunnel Study Group in 1958–60 and 1964–65. Several thousand miles of Sparker survey covered the sea bed, calibrated by 100 borings from floating craft, in which many forms of test were carried out.

Under private enterprise a highly designed tunnelling machine was driving the service tunnel out to sea from Dover in 1975 while the French were completing their initial works. **Figures 24a and 24b**. Then the British Government reverted to type and once again stopped the work which, as in 1882 was moving at fast speed in the uniform dry Lower Chalk.

In contrast to this attempt to join Britain to a whole Continent by rail the Japanese boldly embarked upon a tunnel to join two islands by the Seikan tunnel. This work was pursued in appalling geological conditions following much exploration. Pilot and drainage tunnels help, as well as the reliance placed on the injection work.

Immersed Tunnels

Figure 25 is plotted from a table in Tunnels and Tunnelling, January 1976 which lists 31 immersed tunnels, mostly for roads. There are two rival *Schools*. The American prefers to float circular steel units and weight them with concrete. The Scandinavians prefer rectangular reinforced concrete boxes. In earlier works the units had to be joined by compressed air caissons. Now special jointing devices allow the units to be forced together by the head of water.

The talents needed to build immersed tunnels have little in common with those of *Tunnel Men*. They are, knowledge of seamanship, moorings and dredging supplemented by open excavation for building berths for units which consist of structural work carried out in the open.

Metros, Tubes and Subways

These overlap with rock, soft, immersed and underwater tunnels. The new expedients made Cut-and-Cover work more attractive, even in busy streets, as it is expensive to conduct passengers to low levels. These expedients include mud-flush and bored pile diaphragm walls, steel H piles, ground anchors, ground-water lowering and injections. It became possible to build lateral supporting walls in trench and replace the street upon them, so that traffic can quickly flow again while digging continues under cover with few supports to obstruct mechanical plant. This took place in Stockholm, Tokyo and other cities. (Golder & Seychuk, 1967).

From the late 1950's onwards *metros* quickly multiplied, some as new systems, some as extensions of existing lines. In Paris in 1956 a large programme of new express lines was under way, including ambitious deep stations at Défense. The French held to their preference for double track tunnels even in heavily water-bearing sands. The Metro engineers devised a new logical shield. A steel bulkhead sealed off the face which was to be contained by compressed air. Outside the bulkhead in each compartment a miner was to stand and scratch away the sand in the face by tools passing through glands in the bulkhead. The sand was to fall into the invert where one devoted individual was to assist the sand onto a conveyor belt

FIGURE 24a. *Channel Tunnel Machine — View of Head. (Mott, Hay & Anderson)*

FIGURE 24b. *Channel Tunnel Machine — Rear View. (Mott, Hay & Anderson)*

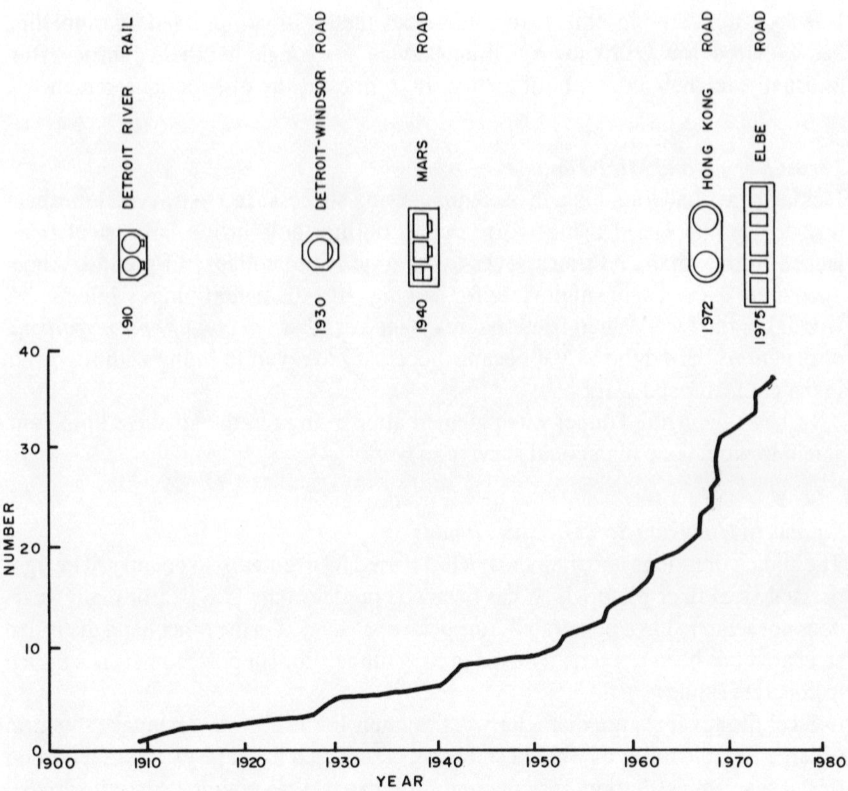

FIGURE 25. *Chart showing increase in major immersed tunnels. (from Culverwell)*

leading to an air lock. In the end the sand had to be chemically solidified by pipes driven from the surface. Air locks were built into the diaphragm so that miners in the end had to work in the compressed air, while the lining behind went on in free air—with difficulties in making it water-tight. The crossing of the River Seine had to be carried out, as in the past, by sinking compressed air caissons.

In London two new lines were started; the Victoria Line (1962–71) and the Fleet Line (1971–), now the Jubilee Line. The interest lies in new linings and the use of tunnelling machines. At two busy cross-roads bored piles were sunk at night, so that later, a pre-fabricated welded steel segmental *umbrella* could be erected on these piles in one week-end. The traffic then drove over them while the booking halls were dug out below.

In the 1960's and 1970's new systems multiplied in cities such as Hamburg, Munich, Vienna, Brussels, Amsterdam, Toronto, Melbourne, San Francisco,

Buenos Aires, among others. In some cities the art of compressed air tunnelling had withered or was unknown so much advice was sought in other countries. But, as usual, each new-comer built up in a short time a cadre of experienced men.

Thrust Boring and Mini-Tunnels

Jacking pipes had long been in use with varying success. In 1950 several mechanical devices appeared using worm-thread boring tools inside large steel tubes jacked from a shaft. At times, over-excavation led to collapse of ground. About 1960 there came a spate of devices for jacking large diameter linings while miners worked in the face. When frictional resistance or *physical conditions or artificial obstructions* barred the way it became necessary to revert to lining with segments in the old fashioned way.

In 1970 the Mini Tunnel was patented after trying out the smallest lining and shield in which one man could survive to work.

Linings in Hard and Soft Ground Tunnels

The choice of linings and their design has varied from country to country. The light pressed-steel liner plate which has been so popular in the U.S.A. for many years does not seem to have penetrated European resistance. On the other hand the rolled steel arch has been universally used in rock tunnels for support, sometimes buried in concrete lining.

Steel linings for large diameter water tunnels in the U.S.A. often take the form of large flexible tubes of welded steel sheets stiffened to shape by internal tension ties. These act as shutters for concrete filling as well as giving a smooth surface. The tunnel has to be completed as it is necessary to work backwards to install each length and weld it to the previous one. Collapses have occurred when the temporary supports have had to stand too long for growing stresses upon them.

In the cases of in situ concrete lining the main advance has been in the use of concrete pumps followed by the pneumatic placer. Steel shutters have become more sophisticated in design and made collapsible with the help of built-in jacks.

Around 1968 Shotcrete came into use so that lining could be sprayed on to rock surfaces immediately for temporary support or for permanent lining. Other variations such as the Bernold system followed in 1970.

The change in linings for soft ground has been considerable. Life was simple for many years when there were two dozen standard diameters of cast iron segments with tables of dimensions and quantities and, as the manufacturers kept the patterns, they were easily obtainable. As the cost of cast iron rocketed concrete began to take over.

In London about 1950 the Metropolitan Water Board found that it would be cheaper to drive an 8 feet diameter tunnel in the clay under London from the

Thames in the south west to the Lea River reservoirs in the north east, than to lay large water mains, even on the surface. In 1951 a new lining, Don-Seg. of solid concrete blocks was used. This was made up of ten identical segments wedge-shaped on the horizontal joints. Alternate segments were placed protruding towards shield jacks opposite each segment. When the shield advanced the jacks drove the segments home and expanded the ring tightly to the clay so that grouting was not needed. With a digger shield progress reached 360 feet in a week.

This success encouraged the M.W.B. to try out a wedge-block lining. This consisted of solid parallel segments to form the ring with a key wedge which expanded the lining to the clay. A variation then was used in the Potters Bar railway tunnel. Heavy interlocking concrete blocks were used but these were wedged out by jacks in pockets left at axis level. A variation was used in France at this period (Donovan H.J. 1968).

A new move now came into fashion in the use of flexible joints in place of the previous rigid joints. In the Victoria Line two new linings were used. One of cast iron, much slenderer in section than of old, dispensed with circle bolts and was jacked out to the clay. The other, of concrete blocks was also expanded from pockets at knee level. Both had flexible joints. In loose ground bolted segments still hold their own.

Many variations in such linings came on the market, especially for sewers and smaller tunnels. Inevitably other types included pre- and post-stressing designs.

In spite of all attempts to achieve complete watertightness the effort was often sensibly abandoned in tunnels housing delicate electrical equipment or in passenger tunnels where droplets could disfigure decor. Inner linings were adopted so that water could be shed to drains below floor level.

The old form of caulking with rust and sal-ammoniac was abandoned. Lead remained popular for use in cast-iron linings under high pressures, but as the price of lead increased, asbestos cement took its place especially in concrete segments and proved effective up to 15 lb per square inch pressure.

Tunnelling Machines

It did not need a genius to suggest the idea of tunnelling by machine, though no doubt it occurred to Leonardo da Vinci. But its multiple development has had to wait until recent years for the new advances in sources of power and modern metallurgy to make it more generally possible. Drinker's masterwork ended during the Critical Decade — 1860–1870, just before Brunel's previous conception of the tunnelling shield at last took off, after lying fallow for so many years. Then followed 100 years of work in which the shield was the basic factor in extending the range of soft ground tunnelling. This much shorter account ends at the time of another notable outburst of tunnelling — in the form of the tunnelling machine of every conceivable, and sometimes inconceivable size, shape and type.

This really *took off* in the 1960's with an even more rapid acceleration in the 1970's.

To record all the recent innovations would require devoted research. This has been provided by the work of Mrs Barbara Stack in Tasmania, who has compiled an Equipment Handbook on *Tunnelling, Drilling and Mining*. This continues Drinker's standard work up to the present day. Nearly half of this comprehensive work is devoted to 90,000 words on tunnelling machines of nearly every type, both for tunnels and mines. As an example of the period of gestation the period from 1846 to 1930 needs over 25,000 words to cover the innumerable patents which far outnumbered the few machines which really worked. Some were actually used or hopefully intended for the Channel Tunnel.

The same space is devoted to modern machines for tunnelling in rock, in which among others, Robbins seems to have set the pace. Many types are listed, emanating from U.S.A., Sweden, Switzerland, Germany and other rocky countries. There have been notable successes and some notably expensive failures as the ground, in all its sudden changes will decide which will survive.

The title of *Shield Excavators* is chosen for machines used in soft ground tunnelling where the mechanisms have to be enclosed in a shield to protect them from the malevolence of the ground.

The build-up to these late developments began in several countries. In Britain the first advance over the earlier types already mentioned was the Kinnear Moodie Drum Digger for use in clays, in 1955. Though still enclosed in a Greathead shield it improved on the Price Digger. The steel drums which carried the cutter arms rotated on a common race wheel and was driven by hydraulic motors, leaving the face open for access and examination. This machine made its name driving tunnels for the Metropolitan Water Board under London from north to south. In modern conditions this proved cheaper than the old method of laying pipes on or below the surface due to the increased speed in tunnelling. This type of machine was soon followed by a variation in the McAlpine digger, in 1956. Both machines were supplied by London Transport to contractors on the Victoria Line.

In the U.S.A. the rock tunnelling machines designed by the Robbins family caught the public eye. Their design was proved on the 30 feet diameter tunnels in limestone and shale under the Oahe Dam in South Dakota from 1956 onwards. These machines towed a double-decked tail or stage which carried the motive power and controls and also the lifting gear which placed rings of circular steel ribs immediately behind the machine and around the tail itself.

Soon many more sophisticated types began to appear and as experience increased the choice and design of cutters became even more decisive than the robustness of the machine and its mechanisms. Efforts were made to bore ever-harder types of rock. As the machines became more complex so did the length of the *tail* to carry the equipment and the operator, especially in small tunnels. Full face rotating arms became challenged by multiple boring heads.

The desire to restrict air pressure in soft ground tunnels to the face behind a bulkhead was proved successful, with some difficulties by the Robbins machine used in the Paris Metro from Place d l'Etoile to the Seine at Neuilly. The ground consisted of water-bearing sand between beds of limestone. The spoil was carried up a tube leading from the diaphragm to an automatic air-lock hopper, untouched by human hand. An ingenious mechanism allowed the driver, while still in his cab, to erect segments one at a time as the shield moved forward. The difficulty, as in so many cases, was the sealing-off of the tail of the machine, behind the lining.

Many attempts were made to maintain a constant pressure at the face in soft ground, ahead of a bulkhead in the shield, including filling the face in various ways under the title of earth balancing machines.

The idea of filling the face of the machine with slurry was patented and successfully solved by the Bartlett Bentonite machine in 1971. The bentonite slurry supports the face and helps to flush out the spoil, as in the drilling of oil wells. The mixture is pumped to the surface where the bentonite is reclaimed. This was quickly followed by other variations and the Japanese soon tried out a number of slurry machines to suit their conditions. (Bartlett J.V. et al 1973).

Throughout the years Patent Agents have prospered from the multiplicity of patents which they have drawn up, but in the absence of mass production it is doubtful how much protection has been afforded or infringements identified, though some inventors may gain credit.

Tunnelling machines are many times more expensive than shields of the same size. Until recently contractors were reluctant to indulge in capital expenditure and relied on labour to dig or bore and blast. The outstanding feature of recent years is the way Contractors and Manufacturers have been driven to invest so much heavy capital expenditure in developing such complex machines.

There is also the heavy cost of proving the hopes of the designers in tackling the ever present geological surprises which lurk below ground. This is a reminder that the only reliable site investigation in any tunnel is posthumous — what is revealed by the completed work.

Success in breeding tunnelling machines breeds more tunnels, while the possibility of fresh needs for tunnels breeds more machines. As a single example, consider the Tunnel & Reservoir Plan (T.A.R.P.) for driving 200 kilometres of tunnels in dolomitic limestone under the City area of Chicago. The purpose is to store the excessive run-off from the City drainage system after heavy rains until the disposal works can treat the effluent at leisure before delivering into long suffering Lake Michigan. As mentioned earlier, the bed of the Lake is already riddled with tunnels to discharge effluent and also to draw in *fresh* water.

A number of these tunnels are driven from the bottom of shafts. This is a helpful reminder that in such cases, in order to make room for the building of the machine and its vital tail, old fashioned miners have to tunnel in the old style for some distance. This is a reminder that the silicon chip will never lead to the complete automation of tunnelling. Many types of machine will be used from mini-size to diameters larger than 10 metres.

Just 100 years ago Drinker ended his monumental work which brought the story of tunnelling up to the work and methods which were the latest in 1879. This history extends to the 100 years which followed and reveals the rate at which tunnelling accelerated and multiplied both for old and unforeseen new purposes. New and changing methods, inventions and expedients developed side by side with tried and old established methods. Figures 26a and 26b.

This Author has lived through all the 20th Century so Part II covers his personal involvement in some of the lessons learned in this evolution.

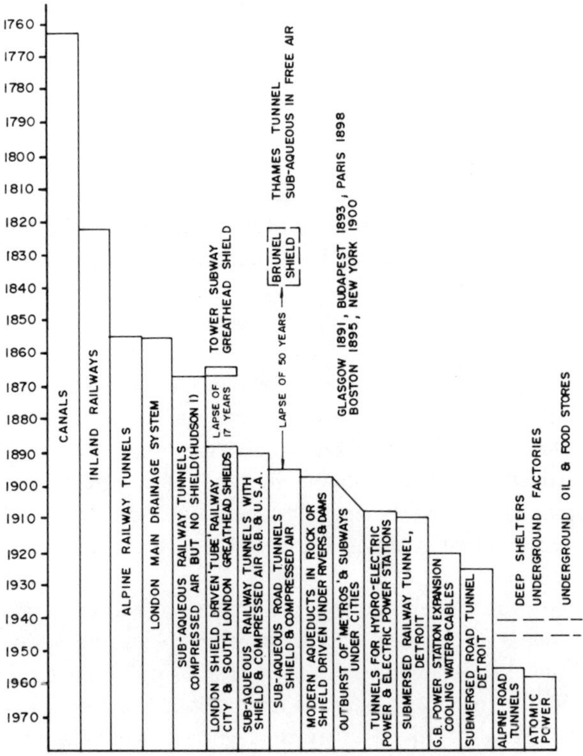

FIGURE 26a. *Chart showing dates of new uses for tunnels.*

	EVENTS	ADVANCES	ROCK TUNNELS	SOFT GROUND	NOTES
1790	(CANAL AGE)		GUNPOWDER SINCE 1681		
1800					
1810		BETTER STEAM PUMPS - CRANES			DAVEY SAFETY LAMP
1820	I.C.E. FOUNDED RAILWAY AGE STARTS	STOCKTON - DARLINGTON RAILWAY PORTLAND CEMENT		BRUNELS PATENTS - SHIELDS	THAMES TUNNEL 1824-42
1830				COCHRANE PATENTS COMPRESSED AIR	RAILWAY TUNNELS G.B.
1840		WORTHINGTON PUMPS			
1850	MILD STEEL PRODUCTION	CENTRIFUGAL PUMPS DEVELOP	COMPRESSED AIR DRILLS (ALPINE TUNNELS)		LONDON MAIN DRAINAGE 1855 - 74
1860	U.S.A. TRANS CONTINENT	HYDRAULIC POWER ADVANCE	DYNAMITE. FREEZING IN MINE SHAFTS STEEL RIBS	GREATHEAD SHIELD PATENTS	LONDON "UNDERGROUND" 1864 GREATHEAD TOWER SUBWAY 1869
1870	REINFORCED CONCRETE PATENTS	GAS ENGINES	BEAUMONT MACHINE DIAMOND DRILLS	COMPRESSED AIR HUDSON (I)	CHANNEL TUNNEL PILOT
1880	ELECTRIC POWER 1ST POWER STATIONS ELECTRIC TRAINS	ARC WELDING DAIMLER PETROL ENGINE ELECTRIC BATTERIES	HYDRAULIC DRILLS	COMPRESSED AIR WITH SHIELD BIG SHIELDS, SARNIA HUDSON	CITY & SOUTH LONDON TUBE 1886
1890		GAS ENGINES DIESELS PATENTS		MEDICAL AIR LOCK HOODED SHIELD CLAY ROCKETING PRICE DIGGER SHIELD 97	NEW YORK RIVER TUNNELS BOSTON SUBWAY 1895 PARIS METRO 1898
1900	FIRST FLIGHTS BY WRIGHTS FORD MOTOR CO. FOUNDED	OXY- ACETYLENE BURN & WELD		STAGE DECOMPRESSION	NEW YORK SUBWAY 1900 PETROL VEHICLES DEVELOP
1910	MOTOR TRANSPORT ADVANCES FIRST WAR, AIRCRAFT DEVELOPS	CATERPILLAR TRACKS ESCALATORS SUBMERSIBLE PUMPS	JACK HAMMERS	TUNNELS IN TRENCH WARFARE	CATERPILLAR TRACTION TANKS - MACHINES
1920	B.B.C. & RADIO TELEPHONE EARLY PASSENGER FLIGHTS	SELF PRIMING PUMPS STEEL SHEET PILING BORED PILES	CEMENTATION	COMPRESSED AIR SPADERS GROUND WATER LOWERING	
1930	WORLD SLUMP CATERPILLAR DIGGING PLANT DEVELOPS	EARLY PLASTICS SOIL MECHANICS DEVELOPING	DETACHABLE BITS CONCRETE PUMPS	JOOSTEN PROCESS BOLTED CONCRETE SEGMENTS	MOSCOW UNDERGROUND 1932 MASS TUNNEL
1940	2ND WAR ATOM BOMB COMMERCIAL FLIGHT COMMON	IMPROVED GEOPHYSICS PLASTICS & RESINS	TUNGSTEN CARBIDE BITS IMPROVED LOADERS EIMCO ETC... CONCRETE PLACERS	SINGLE FLUID PROCESSES ELECTRO OSMOSIS ASBESTOS CEMENT CAULKING CLAY INJECTIONS TUBE-A-MANCHETTE NEW CHEMICAL PROCESSES	UNDERGROUND SHELTERS FACTORIES ETC...
1950	ATOMIC POWER STATIONS MANY LARGER POWER STATIONS CHANNEL TUNNEL SURVEY & REPORT 1958 - 60	DIAPHRAGM WALLS TOWER CRANES MARINE GEOPHYSICS MORE PLASTICS & RESINS	ROOF BOLTING LIGHT AIR-DRILLS DEEP SHAFT SINKING OAHE ROBBINS MACHINES	DON - SEG LINING K·M DRUM - DIGGER (55) MWB EXPANDING LINING FLEXIBLE JOINTS PRE-STRESSED CONCRETE LINING	ALPINE ROAD TUNNELS START UP
1960	MORE TRANSISTORIS- ATION CHANNEL TUNNEL 2ND SURVEY 1964 - 65 INCREASE IN PUMPED STORAGE MOON LANDINGS MORE COMPUTERS	LARGE BORED PILES MORE THRUST BORING LASERS SHOTCRETE	LADDER DRILLING ROCK MECHANICS MORE TUNNEL MACHINES IMPROVED CUTTERS SHOTCRETE BOOM - CUTTER	VICTORIA LINE NEW LININGS IRON & CONCRETE FREEZING IN TUNNELS MINI - TUNNEL	MANY NEW CITY SUBWAYS INCREASE IN HYDRO - ELECTRIC INCREASE IN IMMERSED TUNNELS
1970	CHANNEL TUNNEL PILOT TUNNEL STARTS THEN STOPPED	MINIATURE T.V.	HYDRAULIC DRILLS REVIVED BERNOLD LINING STILL MORE TUNNELLING	BENTONITE DIGGER ROAD HEADERS MACHINES OF MANY KINDS	

FIGURE 26b. *Chart showing dates of major events, expedients, inventions and new methods.*

AN INTERLUDE

HOW SOIL MECHANICS CAME TO BRITAIN

One of the turning points in the History of Civil Engineering is the conversion of the Profession to the Gospel of Soil Mechanics of which Doctor Karl Terzaghi was the true Messiah, aided by his multinational disciples among whom the British played their part, among the backward natives in Victoria Street.

In the Oratorio, which celebrates the missionary years in Great Britain I played the continuo among the Solos, Duets and Trios of our Prima Donnas. Great events from trivial causes spring and some of these causes are hinted at in previous pages, which led to my involvement.

The first was in 1932 when John Mowlem & Co. made their agreement with Siemens-Bau-Union to use their patents for the Joosten Process and Ground-Water Lowering with myself as the nominated expert. In this story the Building Research Station plays an important part and in this year I became involved with it.

Sir George Burt, the Head of Mowlems was a keen supporter who helped to bring about the creation of the Station and he became chairman of the Building Research Board. He sent me to meet the Deputy Director, Dr Glanville (later Sir William and President of the Institution of Civil Engineers) for him to make tests on specimens of Joosten samples to prove its longevity. Later he persuaded Sir William to research into piling problems which had arisen at Fords, Dagenham.

It so happened that in 1933 the B.R.S. started a Soil Physics Section to continue the research into lateral earth pressures which had been started by Prof. C.F. Jenkin. The British Association Committee on Earth Pressure voted £10 towards the cost. Under the eyes of Dr Glanville, Dr Cooling, a physicist, took up the work with two young civil engineers, Golder and Skempton.

Another trivial cause arose when in 1935 Wynne-Edwards was engaged as Deputy Agent on a contract that Mowlems had just obtained to build a reservoir for the Metropolitan Water Board at Chingford. I found him sitting in the drawing office at the Head Office where he had been sitting for a week and no one had spoken to him. I went up to him and said my name was Harding. He brightened up and said Wynne-Edwards. My reflex reply was "Why, you wrote that paper on piles in Vancouver. Come and have lunch". This led to a friendship which became very close indeed when we shared a flat in Chelsea for the last three years of the War as Grass Widowers as our families were out of London. He plays a vital part of this story and later he too became President of the Institution, 1964/65, and knighted as First Chairman of the Council of Engineering Institutions.

Then in 1936 when I was appointed as Agent on the contract for the Central Line extension from Bow Road to Leyton I was able to arrange for R. Glossop to rejoin the firm to take over the specialist work, of which a great deal was ahead. He had succeeded me to set out Leicester Square Station when I was sent to Dagenham and had then spent three years on the Gold Coast (Ghana) starting up a gold mine in the back of beyond. He was an old friend, a mining engineer who was at the Royal School of Mines while I was at the City & Guilds College.

At last there were two like-minded colleagues who could combine their forces to bring about far-reaching results. Incidentally when I was passing over to him my hard won experiences we went to the I.C.E. Library to see what was the latest work on Ground-Water Lowering. The only reference in the massive index of international articles was *Ground-Water Lowering and Chemical Consolidation of Foundations, Work carried out at Messrs. Bentalls Extension, Kingston-on-Thames*, by H.J.B. Harding (Concrete and Construction Engineering).

As a great deal of injection work was specified on my tunnel as well as others we were able to use my main working site and office on both sides of the Pudding Mill River at the end of Marshgate Lane, Stratford, as a base away from the firm's normal depots.

At the end of 1937 the bank of Chingford Reservoir collapsed. The firm had wanted to be allowed to remove the bed of soft recent clay above the gravels of the Lea River but the M.W.B. had refused. The young men at the B.R.S jumped at the chance to analyse a full scale failure. They produced a solution which was derided by the M.W.B. Engineer and by some Victoria Street Consultants (L.F. Cooling and H.Q. Golder, *The Analysis of the Failure of an Earth Dam During Construction*, J. Inst. Civ. Eng., Vol. 19, November 1942).

About this time Wynne-Edwards had taken over as Agent at Chingford and Glossop had joined him as deputy, returning the specialist work to me. Things looked very desperate for Mowlems as they were blamed for the occurrence. It was the first time that the newly developed tractor and scraper plant had been used in this country, leading to rapid bank construction. But Wynne-Edwards was better informed than any one in the firm, having worked in Canada and the U.S.A. and a member of the American Society of Civil Engineers. He said, "We must get hold of Dr Terzaghi". He traced him to Paris through another *trivial cause* because it was Siemens who helped and gave Wynne-Edwards a letter of introduction to take with him on his flight to Paris. So our agreement with Siemens yielded unexpected fruits.

Terzaghi came, supported Cooling, Golder and Skempton whole-heartedly and proved that they were right. The slip passed through the layer of soft clay which should have been removed. Terzaghi produced a revised design for the bank which was to be based on measured values of the strength of the soils as the work proceeded. In order to do this it fell to Glossop to organise boreholes and sampling over the site but, more importantly to set up the first soil mechanics laboratory in this country which was not for research but can be called *Commercial*.

"This occurrence and the confidence which it gave to those interested in the subject, provided an impetus to the development of soil mechanics. Universities, Railway Companies and Government Departments, and even private firms sent selected engineers to the Station for instructional purposes. The Station was then overloaded with soil work, and these men were put to work to assist the permanent staff, so that they were grounded in the subject by that valuable method-practical experience". (James Forrest Lecture – Harding 1952)

Naturally Glossop was foremost among these men, and because of the close association over Chingford he became very much involved with Cooling, Golder and Skempton, and they became firm friends. On the Mowlem side he was closely involved with Wynne-Edwards and me so there were two trios who became entangled in this story. Terzaghi took a great liking to Wynne-Edwards' strong personality and when ever he visited this country, after the War, insisted on staying with Wynne-Edwards in his house in the country outside London, rather than at the Ritz or Savoy. As Wynne-Edwards and I were working very closely together in matters in the Institution, as well as being so close during the war, I found myself accepted by Terzaghi as a personal friend and correspondent.

A cherished possession is a spoof imitation of one of the Harvard Soil Mechanics Series entitled *From Theory to Theory in Soil Mechanics* written by some of his most distinguished former pupils. Terzaghi sent it to me signed by his wife and himself inscribed "To my good old friend Harding, to keep him up-to-date!" Not long before he died I was in the U.S.A. and my wife and I were able to fly to Boston just to have tea with them both, which was a pilgrimage well worth taking.

But back to our story. Terzaghi became much in demand in England once Wynne-Edwards had enticed him over. He gave a series of lectures at the Imperial College and then in 1939 delivered his James Forrest Lecture *Soil Mechanics — A New Chapter in Engineering Science*. This proved a most exciting occasion.

These two years waiting for the war to start proved fruitful. Dr Glanville had been transferred to the Roads Research Laboratory where he extended the British practice to cover soil problems in road design and construction to good effect. Glossop and I had started publishing articles and joining in discussion at the Institution of Civil Engineers which had begun to be alerted.

"The words *Soil Mechanics* first appear in the Index of the Institution publications when a report appeared in the Journal for November 1936, of the First International Conference on Soil Mechanics and Foundation Engineering held at Harvard, U.S.A. Two hundred delegates had assembled from twenty nations, which shows how much interest had been stimulated in other countries. Only one British delegate attended and two British Papers were presented". (James Forrest Lecture). The solitary British delegate was Dr Cooling.

In 1940 when the war started in earnest things began to hot up in many directions. Wynne-Edwards, after some initial exploits was lent by John Mowlem & Co. to spend the rest of the war as Director of Plant in the Ministry of Works. Our friends in the Building Research Station became involved in much full scale work on war damage, and on earthwork failures at Ordnance Factories, airfields and other constructions.

Glossop and I became involved in many changing activities. Among other affairs he built airfields, including experiments in soil stabilisation. After I had spent many months repairing bomb damage to tunnels and sewers in the East End of London and the docks, not forgetting Greathead's Tower Subway, followed by some *classified* tunnelling, I still remained in London on much else.

All six of us, the Mowlem trio and the B.R.S. trio contrived to keep meeting socially to discuss our interests, sometimes staying at Watford to be near the Station. The researchers would describe their work and expound theory while we were able to quote chapter and verse from experiences which supported their notions and fortified them.

In spite of all the restrictions the I.C.E. became more active than ever once the first bombing was over. Possibly because there was not much for those of us in the London area to amuse us. "During the period of the war the Institution of Civil Engineers published more than thirty papers on Soil Mechanics subjects. These papers were welcomed by workers in this field. New definitions had to be learned and a description of strength tests, which the average engineer could not then readily obtain, made them appear academic at first, but as facilities for testing multiplied, so the value of these papers became better appreciated" (Harding 1952).

The facilities for testing at our little laboratory at Chingford were kept fully operative by Glossop's assistants, while much injection work went on where needed. Later a shallow well system lowered the water in an area in the Surrey Docks for the building of Phoenix units for the Mulberry Harbour.

Another activity during the war was the formation of the Works Construction Division of the I.C.E. which became very active and popular. Other divisions also were active such as the Road and Airport Divisions, in which we became vocal, including the growing number of others who were attracted to the Faith. These eventually became so popular with members that the President persuaded the Council to severely emasculate them as they drained away the audiences for the formal meetings. Unfortunately paper was rationed during their most productive period so reprints of papers were only sent to those who asked for them.

The Secretary in the wartime period was E. Graham Clarke who, in spite of a wrong impression among members that he was against change, was very active in promoting these activities and deserves much credit.

The Site Investigation Code of Practice started in 1943 and many of the men I have mentioned served on this code and the later ones on Foundations, Earthworks and Earth Retaining Structures.

My own first entry into an Institution Committee was to advise the War Damage Commission on War Damage due to Earth Movement, about 1941.

This was difficult to assess so I submitted a table on which I listed every form of peace-time earth movement which I could think of and its possible effects on various structures. The idea was accepted that this could be used by the Commission to eliminate the many false claims which were beginning to come in.

But another eventful meeting must be recorded. During the blitz I spent a night with Glossop in his caravan at Chingford Reservoir, and during the nightly cacophony we drew up on the equivalent of one sheet of foolscap a Prospectus for a subsidiary firm which could bring together an expansion of the most modern forms of Site Investigation with Soil Mechanics Laboratory and supporting experts and also future development of every special process or expedient that was possible. Among the imaginary cast-list the name, unknown to him, of Dr Golder appeared in a Principal Part.

This idea was duly fed to Sir George Burt but nothing more was heard during the general controlled confusion. Then in 1944, when Glossop and I had completed our Phoenix Units the Secretary of Mowlems asked me "What are we to do with this new company the Directors have registered as Soil Mechanics Ltd?" This came as a surprise as they had forgotten to tell us. I have always preached that the Tactical Handling of Company Directors is an advanced Science. We just took it over and then told them what we were doing.

Hugh Golder had joined the main firm two years before to qualify for his Associate Membership, as it then was, by practical experience on a contract. This was the start of twelve fruitful years of most harmonious collaboration. At first we were rather shocked at the wholesale choice of title but this was the very first firm in this country to practice the science on a combined professional and commercial basis. The combination of Glossop, Golder and Harding made itself felt.

Although two of us could muster a sound experience in varied civil engineering work and mining, the secret of the rapid success of Soil Mechanics Ltd. was the work of Dr Golder. He started a completely new initiative by taking young Geologists onto our payroll so that every sample from bores in the field could be accurately described at the Laboratory and not left to the *One intelligent and two unintelligent labourers* which had been the custom of well boring firms before the war.

Apart from his work in organising and developing the leading soil mechanics laboratory in this country, he engaged and trained young engineers who soon became competent. Consultants and Contractors were much impressed to receive bound reports with properly drawn-out borehole logs, clear test results and firmly expressed conclusions. This was a real novelty.

From the start we insisted on a high standard of professionalism although the firm had to be, of necessity, *commercial*. The British Consulting Fraternity, over the years had so evolved that they were in no position to undertake the employment of Labour, Plant and the running of Depots without endangering their *Amateur Status*.

Year by year specialist firms began to appear who all seemed to model themselves on our standards, of which a number had been adopted by the Code of Practice of Site Invesigation. These newcomers were able to realise the truth of one of Glossop's aphorisms, "Is it not interesting? You have only to put up a brass plate and call yourself an Expert and you at once begin to accumulate Experience".

During the war I had maintained a nucleus organisation on my working site at Marshgate Lane from which I carried out the many minor works which I handled. So, with a stroke of a pen, I turned them, clerks, timekeepers, electricians, fitters and carpenters into Soil Mechanics Ltd. and used this as our depot for some years. We gradually evolved laboratories in Victoria Street and later in Old Church Street, Chelsea, and the firm grew apace.

By this time other voices were heard, but in 1945, Four Lectures on Soil Mechanics were arranged by the Institution, three of the authors being Cooling, Skempton and Glossop. Many papers about that time were written by permutations and combinations of the six men that I have grouped together. Mine were more angled to the purely practical side with the Works Construction Division, leaving theory to the others, though I did announce the Harding Coefficient of So-Whatness to be applied to theory when it went beyond the limits of common sense. I also heard myself say in the discussion of a paper by Hugh Golder and myself to the British Association that "Nature abhors a Differential Equation".

Wynne-Edwards became a Director of Costains at the end of the war and started the second firm in this country led by Serota, who trained briefly in our laboratory by friendly arrangement on his release from the Air Force. One of the pleasant features of all this story of friendly combination and rivalry is the lack of that spiteful hostility which seems to afflict the Pure Scientists.

We must not forget the publication of Géotechnique which for two years was edited and produced by Glossop and Golder, and then taken over by the Institution. The Geotechnical Society also got going under the same auspices. One innovation was the word Geotechnology to cover much and this leads to a poem by Dr Golder:

> Impassioned eloquence arise
> Salute the Engineering Dawn
> Manna descended from the skies
> When 'Geotechnology' was born.
> Cooling, Skempton and Glossop cast
> Their pearls to serve a porcine meal,
> What matter if the pearls be false
> So that the swine be real.

A final quotation from my *James Forrest Lecture* shows the rapid growth of interest in the subject:

"The Second International Conference on Soil Mechanics and Foundation

Engineering was held at The Hague in 1948. There were 596 participants from thirty-eight countries. There were sixty-nine British Delegates and sixty-seven British Papers published. Thirteen British Laboratories presented reports on their organisations and equipment".

The people whom I have mentioned can claim to be as far as Great Britain is concerned, the First of the Many.

Professor Skempton went from strength to strength. He became the first to hold a chair as Professor in the subject in a British University. But more important, he was chosen to succeed the great Dr Terzaghi as President of the International Society on Soil Mechanics and Foundation Engineering.

In this story of the slight self-deification of principal actors, we must not forget the debt to our faithful operative foremen on borings and processes whom we trained at the same time as we trained ourselves.

One particular case comes to mind. I have already described well-point work to help the Royal Engineers in their bomb disposal work. We were soon called in by the Admiralty who had a section whose work was to dig out and de-fuse unexploded mines which the German Navy had given to the Luftwaffe to extend their unpleasant attentions to us.

One of our foremen, Birkett, was so successful in extracting a mine which had passed into soft ground near Yarmouth by means of an installation of our well points that the Admiral in charge insisted upon his being given a commission. At first the Admiralty objected on snobbish grounds as Birkett had only a Local Authority education.

Lt. Birkett did so well, and developed a system of extracting mines which fell in soft river mud below tide level. This was by means of a contraption of three rings of pipes with many jets which could face downwards, upwards and horizontally. This was lowered from a small barge containing three trailer fire pumps and could jet a hole down to the mine by changes in the choice of jets. A number of interlocking blocks were cast which could be built up during jetting to form a primitive caisson which sank as the units were added and the mud jetted out. Birkett was decorated for his work.

The very first Site Investigation carried out by Soil Mechanics Ltd. in 1944–45 was an extensive one in H.M.S. Vernon the Portsmouth Torpedo School from which the mine disposal squads were based. One of many cherished memories is of being entertained in H.M.S. Vernon Officers' Mess by Sub-Lieut. Birkett, R.N.V.R., M.B.E.. I felt very proud and slightly humble. (Hunter, J.W., 1948).

Part II: TUNNELLING EXPERIENCE

INTRODUCTION

Every Civil Engineer, however academically endowed, has to start his practical career in a state of ignorance and then gradually accumulate experience. The lessons from the past are not always out of date so an intelligent interest in the antics of his predecessors can speed up the process.

Books on tunnelling multiply, sometimes by applying *scissors and paper* to the works of previous authors. Most books concentrate on long important tunnels which had powerful financial resources. This gives small comfort to many of us as most troubles arise in short tunnels of intricate twists and grades where resources, financial or otherwise are small.

Field Marshall Sir John Burgoyne, a famous Royal Engineer, who helped to found the Institution of Civil Engineers of Ireland while he was Chairman of the Irish Board of Public Works, wrote in the 1830's that the dissemination of Civil Engineering knowledge was best carried out by practical experience and personal intercourse.

The lessons of practical experience can also be disseminated to others by the written word in books, articles and papers to learned Institutions.

The lessons of 55 years experience which follow change their shape due to a change in career. I first spent 34 years with the contractor, John Mowlem & Co. Ltd. from junior engineer to Company Director, which included the time with Soil Mechanics Ltd. from its founding. Then from 1956 I worked as a single handed Consulting Engineer with an International Practice, in which tunnelling only played part, compared with many other diversifications. Both these periods covered many forms of work other than tunnels. Work in the open air by sea or land provided quite different experiences and affected the mental approach to tunnelling problems.

During and after the second war some of the most valuable papers to the Institution of Civil Engineers were those to the Works Construction Division. But due to the continued rationing of paper, copies were only distributed to members who asked for them so they are not so easily found as those in the normal Proceedings.

Among these are two which I was asked to write for the Division. *Site Investigation including Boring and Methods of Sub-surface Exploration* (Harding 1949) was to introduce the first Code of Practice on Site Investigation which was a novelty to the profession.

The Choice of Expedients in Civil Engineering Construction (Harding 1947) embodied my personal philosophy, learned from experience. Many advances have been made since then but my opening paragraph still stands: "In most civil engineering works worthy of the name the unexpected happens; to be prepared for such eventualities and to forestall their effects is the test of good constructional practice".

This applied particularly to tunnelling which is completely earthbound, as the tunnel engineer may often be surprised but should never be astonished.

There are three main causes of the unexpected:

(a) The ground in its infinite variety and its water content
(b) The behaviour of plant and equipment
(c) The behaviour of human beings. The stupidity or cupidity of individuals or organisations, governmental or otherwise can stultify both theory and practice.

Many of the events which have come to mind and are described naturally go back into the past. This past may seem irrelevant to engineers starting their careers with up-to-date plant and materials. Though much has been modernised much remains the same especially *in the Earth beneath and the waters under the Earth*. But, among much else described, warnings of incidents, failures and collapses especially in brickwork can be useful, for most British cities and towns are riddled with elderly brick-lined tunnels for rail and sewer, a number of them empty and deserted. These from time to time trigger off a collapse in a street and so form unsuspected elephant traps for the unwary. Alertness can often prevent or minimise damage.

The rate of gaining experience can depend upon intensity as well as time, and with time comes the increase in personal responsibilities. My most intensive and formative experience was gained in soft ground, so the first chapter sets the stage which led by chance to intensive work on geotechnical processes, before they became dignified under the title of Geotechnology. These come in the second chapter and after that the subjects are grouped under headings.

Very long tunnels can be monotonous and provide little variety of experience for a young engineer. So emphasis has been placed on more complex works. These may have less *News Value* but they have provided much variety.

Chapter 6

TUNNELLING IN SOFT GROUND

EXPERIENCE IN THE LONDON BASIN

The world-wide development of Shield tunnelling with cast iron linings derived from the intense activities in the densely populated London area. The geology of the London Basin provided variety of experience contrary to the popular idea that the whole area is a bed of uniform clay.

A series of Eocene beds were laid down upon the chalk which was then forced into a syncline under London with the chalk outcropping on the North and South Downs, and the Chilterns. **Figure 27**.

Erosion was assisted by the River Thames flowing from west to east. It meandered widely west of London and deposited vast areas of water-bearing sands and gravels which make an ideal concrete aggregate, of which a vast wealth is sterilised under London Airport.

Then the land and the water level of the North Sea fell in three spasms. The Thames cut down its bed between the clay hills of London and laid down three terraces of sands and gravel. It also caused the top of the London Clay to become a very uneven eroded surface below these terraces, with random deep holes whose origin is still debated. All of these are filled with water-bearing sands and gravel.

Immediately east of the City the river became a wide estuary which allowed 40 feet or more of soft estuarine clay to be deposited over the gravels.

To add variety, there was also a small anticline running north and south along which the River Lea has carved out its valley and which brought the Woolwich and Reading varied beds to the surface. These were then eroded away by the River Lea depositing its own gravels upon it. This provided Sir Marc Brunel with all his troubles and the writer with much later experience (Rolt 1957).

The Woolwich and Reading beds provide even stiffer mottled clays mixed with random varieties of sands, clays, cobbles and thin beds of limestone before the fine water-bearing and difficult Thanet sands are met below them, resting on the chalk. All these outcrop in South-East London which partly explains why the London Underground has fought shy of this area on the ground of cost. Drainage engineers have had no choice and many a small contractor has ceased to exist, baffled by ground conditions.

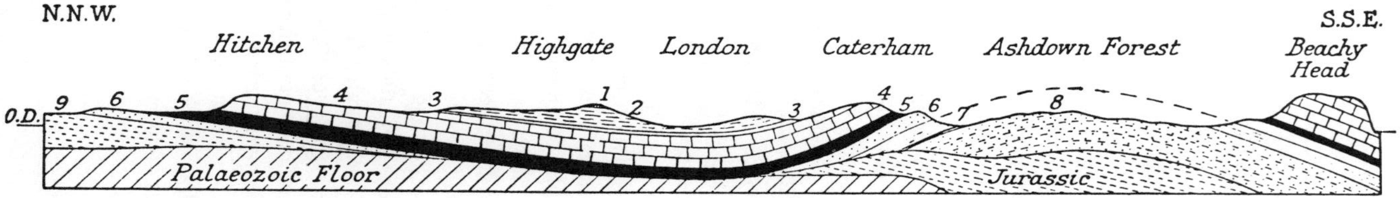

Showing the Palæozoic floor beneath London. Length of section approximately 100 miles ; vertical scale exaggerated about 20 times. (1) Bagshot Sands. (2) London Clay. (3) Woolwich and Reading Beds and Thanet Sand. (4) Chalk. (5) Gault and Upper Greensand (the latter is not continuous under London). (6) Lower Greensand. (7) Weald Clay. (8) Hastings Sands. (9) Jurassic rocks.

FIGURE 27. *Geological section across London Basin. (Blythe)*

This note explains the variety which can be met, and why Brunel's preference for iron lining proved itself, while the old timber methods still had scope.

The London Basin provides much varied experience. Shafts must be sunk through water-bearings sands in the various terraces, to drive tunnels at depths beneath them, while shallow tunnels meet the same variations. Today special processes solve the problems (at a cost) which were previously dealt with by slow, risky and eleborate methods of tunnelling. The Woolwich and Reading beds are always lurking to catch the unwary.

The text books describe the working of the Greathead shield, whether plain or hooded, as this ground is ideal for such work. But before such shields can be installed at depth, much shaft sinking and hand tunnelling is needed which can call for skills, especially in *breakups* for chambers and constructing openings, which are not often described. It would need a book on the lines of *Underpinning* by Prentis and White to give details of the ingenuities which should be covered. Bolted segments are best suited for this type of work where segments have to be removed or shortened and jambs and lintels installed.

WORK IN LONDON CLAY

Short lengths of tunnel in clay can make the cost of a shield uneconomic as hand tunnelling is simple but slower in such temporarily self-supporting ground. Tube tunnels in this area vary from *small* of, say, twelve feet diameter downwards to *large*, from twenty-one to thirty feet for stations, escalators or chambers. This generalisation excludes the sub-aqueous road tunnels.

Supporting the face in a twelve foot tunnel is simple enough. The high cost of a station shield becomes justified for relatively short drives of 300 feet as it not only provides erectors for the heavy segments as well as working stages on its cross-members but it especially provides excellent means for supporting the heavy clay face. In addition to these heavy pressures, clay blocks can slip out along the *backs* or fissures peculiar to this clay, unless quickly supported. Large blocks have produced fatalities.

Between the wars a programme of revolutionising stations with below-ground booking halls and multiple escalator tunnels led to the need for large chambers of 26 feet diameter, too short to justify shields. For these the pilot tunnel became invaluable. These worked up from 6 feet to 12 feet and sometimes to 16 or 18 feet diameter. Thus the clay face was never vast and only the area between each diameter had to be supported. This was made easier by using bolt-holes to help in providing fixings for timber. But when at the ends, a full face had to be timbered, massive soldiers, struts and walings had to be handled in confined spaces, before structural steel eased this problem.

When I decided to be a civil engineer at the age of ten I was given the three bound volumes of Engineering Wonders of the World which I absorbed during every holiday. I was particularly interested in one part — *The Tube Railways of London,*

Their Design and Construction by Harley H. Dalrymple Hay, M.I.C.E. although I did not live in London and had never travelled by Tube. This led me in 1911, on Exmouth sands to try to drive a shield tunnel, when lying flat on the sands, using a Bath Oliver biscuit tin, with the end cut out as a shield, but my eleven year old thumbs proved little use as jacks to overcome the friction.

CITY & SOUTH LONDON TUBE EXTENSION, EUSTON-CAMDEN TOWN

Personal experience started in 1922 on leaving the City & Guilds College. In those days of slump the first need was to find an opening and then ask what it was. On being taken on by the contractors, John Mowlem & Co., I found that I was to work on the expansion of the City & South London Railway from Euston to Camden Town. The Engineer was no less than (Sir) Harley H. Dalrymple Hay. My chief, Mr Rowell, had worked on the original City & South London tube railway and later had been Agent on the Waterloo & City Railway. On this he worked closely with Dalrymple Hay when the latter introduced his hooded shield and clay pocket method of dealing with sands and gravel. Before this invention the method was to drive a closely timbered chamber ahead into which the shield could be pushed. So I learned the orthodox drill of the time under the two most experienced men of their day. I had to control seven shields and several hand tunnels from two working sites.

In the Camden Town-Euston extension the cast iron lining was 12 feet 6 inches outside diameter but 11 feet 8¼ inches inside, for reasons never divulged but a strain on arithmetic. In later years the inside diameter was increased to 12 feet or 12 feet 6 inches to allow for faster trains. By 1922 the design and manufacture of Greathead shields had become almost a standardised monopoly of Markhams of Chesterfield. The hydraulic jacks were still water-filled as the simpler oil-filled mechanisms lay in the future.

Salesmen were in evidence peddling the earliest forms of pneumatic *spaders* and meeting with the customary resistance as the miner disliked the noise and the weight of the earlier models. Miners tried to break off too large lumps of clay at a time which caused the early form of blade to break. The miners were used to chipping off thin flakes of the hard clay by quick blows of their short-handled picks and at first kept up with the output of spaders.

Twelve men worked in each shield. Two miners and two miners labourers drove a six foot high box heading far enough ahead of the shield to allow the other miners, after they had dug out the length of a ring, to attach steel joists called piles horizontally from the diaphragm of the shield to the face. **Figure 28**. Two settings of box heading timber would be struck and, as the shield was *shoved* the piles broke off large lumps of clay to help with the next length. (A few years later, when spaders had become accepted, a rival set the pace by abandoning the box heading and so reducing the gang to eight men.) Our gangs were all on piecework and

FIGURE 28. *City & South London Railway extension, Euston to Camden Town, 1923 — Greathead shield with timber box heading. (London Transport)*

always left for home after they had built and grouted four rings without waiting till the end of the 12 hour shift, day and night.

Naturally the cost of the work depended upon the piecework rate agreed upon. The Agent on my first job kept all financial transactions a deep secret so it took a little time to solve these mysteries.

The Civil Engineering Contruction Conciliation Board had made their first Working Rule Agreement in March 1921, a year before my advent. This rule fixed the basic rate of pay of a General Labourer — the *X* Rate — which was to vary with the cost of living. This ushered in a long period of peaceful co-existence. Several appropriate Trades Unions sat on the Board but very few, if any of our men belonged to one, except for Tradesmen. Fifty years later things had changed. From 1922 until at least 1938 *X* remained unchanged except for one year when it dropped by ½d. This shows that, contrary to today's notions, inflation is not compulsory every year.

The miners rate was $X + 6$d. By the 1972 agreement it had changed. *X* was 56 new p. as the decimal confidence trick came in on Decimal Day, February 15th 1971 and the miners rate was *X* plus 7p. The agreement also forecast for 1973 — $X = 57$p and for 1974 $X = 62$p. or in old money a miner's hourly daywork rate would be 12/5d plus 1/5d. Today there are many fringe benefits, unheard of in 1922, together with greatly forced-up piecework rates, so comparisons beyond these would be more odious than usual. One benefit of becoming a Consultant is being no longer concerned with such mundane matters.

On the drives to Euston the unexpected did happen when water-bearing sands in the Woolwich and Reading Beds were met several hundred yards before they had been anticipated. Simple air locks were built with five feet thick walls, the compressed air plant which had been waiting for much later use was put into action, and the sands slowly negotiated and left behind. It was difficult to see the methods as the miners never took their eyes off the face while exposing each bit and the General foreman, a sergeant-major type, tried to discourage beginners. By coincidence, he left soon afterwards. When I asked for the reason I was told "George R was a first class Foreman until he bought himself a set of drawing instruments!".

The sands had been expected while driving round the 300 feet radius curves to enter Euston. The lining was increased to 12 feet 6 inches inside to allow for the *kick* of the coaches as they rounded the curve and two hooded shields had been ordered. These then had to drive in the hard Shepherds Plaid, highly coloured very hard Woolwich and Reading clay, round these sharp curves. Their cast iron cutting edges, which protruded ahead to form the hoods, broke to pieces and the tunnels were finished without the shields *by hand*, as the saying is.

When the main shield drives had been completed and had joined up well within the customary inch, I was put on night work to take charge of the Flying Junctions. These enabled the new City & South London trains to join the existing Hampstead & Highgate Line south of Camden Town Station so that trains to or from the City or the West End could diverge to Hampstead or Highgate. All this work was to be done at night between the normal gap from about 1 a.m. and 5 a.m., with longer shifts on Saturdays and Sundays and without stopping a single train.

At this time the City & South London tube between its terminus at Clapham Common and Moorgate Street was being enlarged at night, to the equivalent of 11 feet 8¼ inches by using the old 10 feet 6 inch segments with four special keys. Large junction and cross-over tunnels were included (Jones and Curry, 1927).

FLYING JUNCTIONS – CAMDEN TOWN

The Flying Junctions at Camden Town and the City & South London widening were the first such works carried out in live running tunnels so we were involved in a new form of work. **Figures 29a and 29b** show the general layout of the junctions. Four *Step plate junctions* had to be built, two on each of the Hampstead lines which were joined up by loop tunnels driven between them but serviced during driving from the completed running tunnels through cross headings. A step plate junction consists of short lengths of ever increasing diameters of cast iron

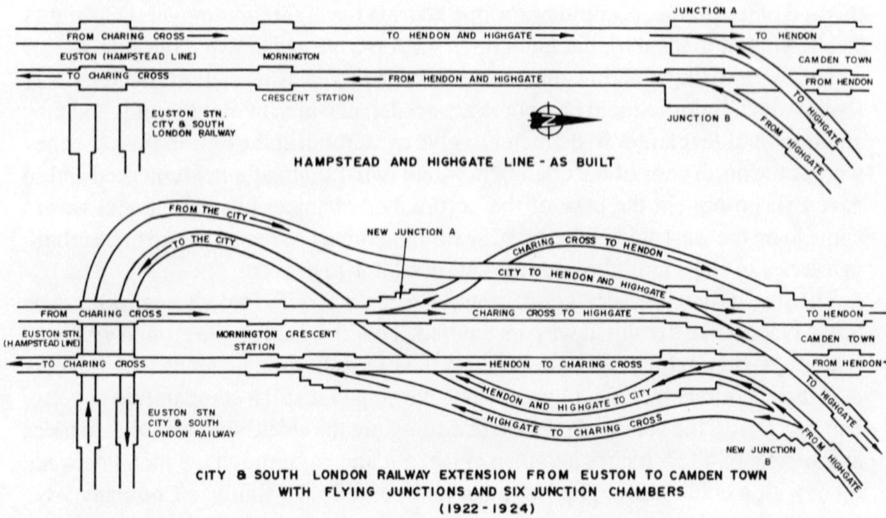

FIGURE 29a. *Plan of flying junctions at Camden Town, 1923.*

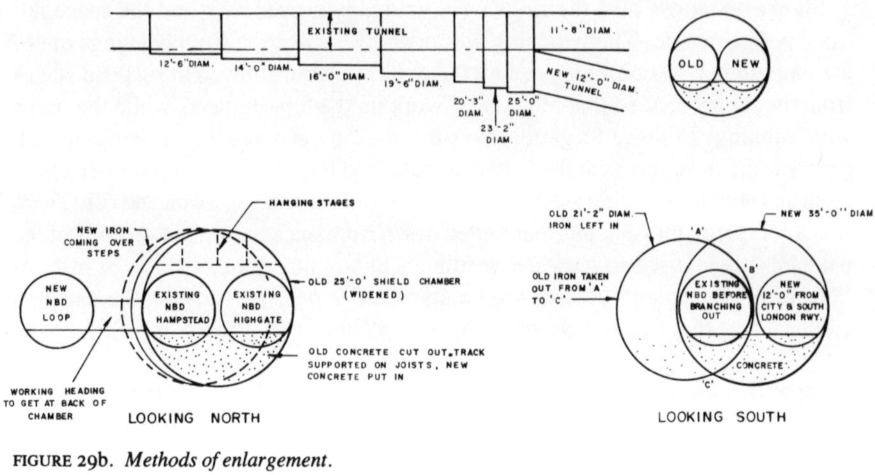

FIGURE 29b. *Methods of enlargement.*

lining built eccentrically round the running tunnel until rings of sufficiently large diameter allow for two running tunnels to emerge at the far end. The twenty-year old rings of the old tunnel were in good condition as was the clay behind them and all voids were found to have been well grouted with blue lias lime grout.

As the new line to the City passed under one of these junction chambers a timbered shaft was driven upwards into a newly widened portion so that spoil from this and other chambers could be dropped during the night for removal. One night I stepped into it in the dark but hung on to the top edge and crawled out safely.

The two existing junction chambers had been driven using station shields and 21 foot 2½ inch lining. Shield chambers are needed in which to build shields and they are of several feet larger in diameter to give extra room at the top. In this case they were at the north ends of the chambers where twin running tunnels emerged and so gave extra room. In the case of the northbound chamber our new tunnel was to come in on the west side of the existing running tunnel so the rings had to be rebuilt in a series of steps until the end rings allowed this to happen.

The shields had been designed, as was the custom, with smooth steel skins to be left in place when the shield was dismantled. This also held up the ground while the final rings were built inside the skin. So the old station shield skins had to be cut out. It is a curious fact that oxyacetylene burning was still a comparative novelty. Until its arrival the standard method of cutting out the shield skin was to drill holes at intervals of a few inches and then chisel out and cut through the metal between them with a cold chisel. The Engineer and possibly the Railway Company were frightened to allow the gas cylinders into the running tunnels so the first skin took several months to cut out. Later courage allowed them to burn out the other which took three days. This drove home the value of new expedients.

Figure 30 shows how the rings were gradually moved over and the space left filled with concrete. The widening was done by excavating from headings out of the loop tunnel. In order to speed up the work we were allowed to suspend stages from the tunnel roof so that men could work on the top segments while the trains were running. In those forgotten days there were open vestibules between each coach as the only access and a Gateman balanced himself between every coach to open and shut low folding steel wicket gates to allow passengers on and off. There was a complaint that odd bits had rattled down from our stage on to the heads of the passengers who used to pack the vestibules to bursting point. I received instructions from my senior with a cartoon marked *not to be kept*, showing the gateman calling "next stop Camden Town — Mind yer Nuts".

TUNNELLING IN LOOSE GROUND *Isle of Dogs Pumping Station 1924–1925*

The clay met with on this first work had been a surprise as nothing so hard or like it is visible on the surface. This work in clay taught me nothing of soft ground. I told my Night Foreman that this worried me and confessed my complete ignorance. He was a famous character, called *Bumper* Harris as he had lost a leg at Waterloo -- while building the Bakerloo Tube. When the first escalators were established, at Earls Court Station to reach the Exhibition, by Mowlems, before the first War, Bumper was given the job of travelling up and down the escalators to inspire courage in the nervous.

When I told him my worry, as I made a point of asking such men to teach me all

FIGURE 30. *Widening original junction chamber at night.*

they could, he remarked, "You know that fellow who was here before you, he was one of those bloody B.Sc's but he didn't know nuffink". Bumper was an untrained artist. A week later he handed me an exercise book in which he had taken the trouble to draw a series of perspective drawings explaining the tricks of timbering in small steps to hold the ground by only exposing very small areas. What small authority is vested in a young engineer is never diminished by asking a good workman for details of how he plies his trade.

Between this and my next job I had to make a record tracing of two old drawings, one of a twin track railway tunnel and the other an eight feet sewer, both brick lined, by the English Method — on bars'. These were in the form of a comic strip — a series of step by step stages in the story. By one more coincidence my next work was to take over and complete a Storm Water Pumping Station in the Isle of Dogs, after the previous contractor had gone into liquidation, blaming the excess zeal of the London County Council engineers.

Much varied work was needed on this contract but the key to the work was the driving of an eight feet diameter bricklined sewer by the English Method, precisely by the methods which I had been drawing. The previous contractor had sunk a sump at either end of this short length, heavily timbered as the tunnel was all in the flood plain gravel and six-inch centrifugal pumps had to work day and night to lower the ground-water by twenty feet.

This was my first of many involvements with the London Main Drainage system in peace and war. One of the benefits of early reading arose from the fact that the system was lucidly and pictorially described in my volumes of Engineering Wonders of the World and I found that I was better informed on many details than some of my elders.

Bazalgette's system of main sewers was designed with a number of weir chambers so that in time of heavy rain, with the vast run-off from such large paved areas, excess flow could be carried by storm water sewers to one of several pumping stations which could deliver this flow through back-flaps into the Thames against the head of high tides.

The tunnel was driven in about ten *lengths* of ten feet. **Figure 31** is a flashlight photograph which I took with a No. 2 folding Brownie camera. The entrance to the bar-heading can be seen behind the bar props and cill props. The bar props hold up the bars until the cill props (which are above the cill) are placed once the cill is in position. Below the cill can be seen the ground props which are placed as the cill is underpinned when the bottom half is dug out. Much was learned by watching and talking to the experienced miners. When an urgent call went up for me to send down *long litter*, I found that I had to send round to a near-by stable for the used straw bedding which contained such a valuable binding agent. Today stables are not so quickly found for such homely materials which have given place to more expensive nostrums.

FIGURE 31. *Isle of Dogs Pumping Station — Sewer tunnel length 'on bars' ready for bricklayers, 1924.*

The photograph also shows the semi-circular *leading frame* which the engineer must set with the top bar to the correct grade. He also had to mark out the point of each course of bricks on the inside of the semi-circle and fix the side boards, which can be seen to mark the axis of the arch. In such loose ground the poling boards must all be left in the ground. The bricklayers worked twenty four hours round the clock to complete a length which had taken the miners a week to excavate. As the brickwork rose, miners in attendance would strike walings and bars as soon as the brickwork had supported the bottom of the poling boards.

As soon as the brickwork reached axis level a timber drum was set consisting of loose ribs and loose laggings. The latter had been wrought to the curved face of the finished lining. A bricklayer worked on either side and added a lagging as required. The miners would strike each bar in turn by lifting to the top of the drum and passing it into the bar heading ahead. Finally two grooved laggings would be placed to leave a gap of a foot or so between them to suit pre-made block laggings. These were short lengths about a brick and half long. A surviving bricklayer then worked himself out backwards into the heading as he added a block lagging at a time. On these he laid the final stretchers end-on to complete the circle, with remarkable accuracy.

For the fifteen feet radius curves on an eight feet diameter barrel the bricks had first to be cut to various lengths. The inner bricks at axis level would be five inches and the outer ones nine inches long. The others had to be adapted so that the perpends could be as strictly kept as the L.C.C. demanded. The Resident Engineer took me down a finished length and said "Just look at that". I said, in my innocence, "Yes, is it not a nice piece of work?" The retort — "it is disgraceful, he has not kept his perpends!" My answer "whatever are perpends?". The reply "when you bend a thin slip of paper round the circle every other joint must be exact on this circle. Some of these are over one-eighth of an inch out". He had the whole inner face cut out (though at the expense of the L.C.C. and only the sewage flow would have seen it.)

When two sewers joined with graceful curves a cut-water had to be built from invert to axis. Nottingham Bricks, used for their softness, had to be cut and rubbed so that each brick had its unique identity and shape. Great precision was demanded so that if two or three bricks were laid in a shift it was considered good, if expensive, progress.

Today cutwaters are made, I am told, in concrete at vastly less cost. I do not think that the sewage minds.

Brickwork was then in its prime and I learned much from the L.C.C. Clerks of Works who had been artists in their time and were only too glad to instruct a younger man.

One Clerk of Works, Sid Field, re-entered my life during World War 2. He inspired a letter which I wrote to the Times about 1968:

ALL CHANGE FOR ST PANCRAS STATION

From Mr. H.J.B. Harding

Sir, — There is another aspect to St Pancras than the silhouette and outward form mentioned by your Architectural Correspondent.

During the air raids I was involved in rebuilding L.C.C. deep sewers in brickwork as and when bombs landed in them, and had to deal with an elderly L.C.C. Clerk of Works who had been a master bricklayer. He was rather a cold and chilly individual and in an attempt to cheer him up I quoted the saying that Sir Gilbert Scott had designed the Foreign Office in Brick Gothic and "with a stroke of the pen had turned it into St Pancras Railway Station".

The effect was instantaneous. The Clerk of Works leapt into life and became quite lyrical with a paean of praise for his craft. He said that St Pancras Hotel was the finest example of bricklaying craftsmanship in existence and rhapsodized over the bonds, king closers, queen closers, rubbers, squints, quoins, noggins and other technical niceties.

The general effect of approaching King's Cross and St Pancras is rather depressing, if one has time to glance around, but an examination of the detailing of St Pancras shows that it is truly a masterpiece of craft even if not always accepted as art. It could be argued that at least part of the structure is worthy of preservation as an example of an almost lost craft of which Sir Gilbert Scott as well as his bricklayers must have been a master in order to design such intricate detail. Yours faithfully, H.J.B. Harding.

This experience was useful when in 1957 I headed a party from the Civils to visit the building of the Brussels Exhibition. We were also shown new motorways being built on the lines of the old City Walls. We were also shown a very complicated work where three roads crossed at different levels, two of them in tunnel, one above the other. A number of distinguished bridge engineers were mystified and I was able to explain that we were seeing the English System being carried out in Belgium by Italian miners.

In much of this story I can claim to becoming involved in several Firsts of new methods, but in this case it is nearing the last. My Foreman on the Central Line Extension told me that in the 1930's he had to build a large chamber on the Mersey Tunnel *on bars* and had great difficulty in rounding up enough miners who knew the drill.

This work at the Isle of Dogs involved what was, unwittingly, a crude form of Ground-Water Lowering system, before such a title had ever been coined. Shafts had been sunk at each end of the drive for the 8 feet diameter tunnel in water bearing Thames sands and gravels. As the water level had to be lowered for twenty feet to the sumps, petrol driven centrifugal pumps had been installed above the original water table, requiring twenty feet long flexible suction pipes. This sytem had been installed by the evicted contractor and kept running day and night by the L.C.C. while a new contract was negotiated.

When we came to drive the tunnel we followed the standard practice of many years by laying an earthenware open jointed subdrain below the invert of each length, not forgetting to pull forward a chain inside the pipe at each length so that any matter apt to block the pipe could be cleaned out. The cone of depression from the front of the pipe reduced the ground-water ahead sufficently for the next length to be driven.

This was before the coming of that blessed invention, the self-priming pump. Continuous crises arose every time the foot valve caught on bits of timber and the pumps lost their water. Then the twenty feet of suction pipe had to be hauled up sufficiently for the valve to be cleared and the pump reprimed in time to prevent the water rising into the unlined length and disturbing its stability. How different from experiments on a clean fixed system in a College Hydraulic Laboratory and how nice it was in later years to pump from filter wells with reliable self-priming pumps.

CHARING CROSS ESCALATORS

Towards the end of 1925 I returned to the West End to take part in a new campaign by the London Electric Railway to modernise and expand their stations. With the revival of tunnel work in 1921 the policy was adopted of gradually replacing lifts with escalators in order to provide a smoother flow of traffic to meet the increase in passengers.

This called for a great deal of capital expenditure for the construction of underground booking halls and circulating areas as well as the large diameter sloping tunnels to carry three escalators with large chambers at top and bottom. In some cases two stages of escalators were needed to reach greater depths with even more chambers. The cost proved itself justified by the increase in revenue as the Tubes were made more attractive. It also provided much appreciated work for the tunnelling firms.

There were lessons to be learned by following up the previous history which led to these expansions. The original Inner Circle Railway, the first of its kind, had been built almost entirely in cut-and-cover with a few lengths of tunnel. The temporary devastation must have been extraordinary to judge from prints and early photographs of these wide heavily timbered cuttings across Parliament Square and elsewhere. Equally remarkable are the steel engravings in early I.C.E. papers which I stumbled upon, in which every board and strut are faithfully shown as the whole length of Cannon Street was dug out for this twin track line. Presumably due to such careful work none of the buildings collapsed.

It must have been a relief to the inhabitants when the chance came to complete the length from Westminster to Blackfriars in the reclaimed area behind Bazalgette's other work of genius — the Victoria Embankment.

Thirty years later the Tubes became popular as their construction was almost invisible except for working sites at future stations mostly off the street.

In 1906 the *Bakerloo* opened its twin station tunnels under the west end of the Charing Cross (Embankment) Station of the District Railway. These were in London clay. Since then this station has undergone a number of surgical operations over many years. A number of these were at the hands of Dalrymple Hay and John Mowlem & Co. The Hampstead line had originally ended at the Strand under the main line station of the South Eastern & Chatham Railway. Then in 1912 it was extended by driving a loop in compressed air under the River Thames and back to form a single station beside the twin Bakerloo tunnels, and so save the need for trains to reverse.

As part of this contract my firm also drove small single escalator tunnels from new interchange subways below the District railway platforms to serve these tube stations and so eliminate the lifts. These were driven through the water-bearing flood plain gravels and the ground-water was lowered by pumping from several old fashioned timbered shafts. I was later told that Dalrymple Hay disarmed criticism from local owners by arranging for the discharge of water to be recirculated below ground by means of adjacent shafts.

By 1925 a firm of the well known contractor Sir John Norton Griffiths had extended the Hampstead Line towards Kennington under the river by cutting through the old loop tunnel. This entailed the need for a second station tunnel on the southbound line and of necessity this was beyond the east end of the District Station. This required more extensive access works and provided me with fresh experience, as a change from the Isle of Dogs.

As so much is written about long modern mechanically driven tunnels it is as well to remember that small complex tunnelling will never be eliminated. While the District Railway engineers picked up the track at nights and concreted in closely centred steel joists before replacing it, we sank a 12 feet shaft in Embankment Gardens and attacked through the end headwall of the old station.

This led us under the westbound platform which was founded upon a dished invert of several rings of brickwork which kept out a few feet depth of the ground-water. The next step after burrowing across to the opposite platform, was to sink a timbered shaft under each platform and pump down the ground-water. This time we boldly pumped the water into the sewer from the ladies convenience, while we cut out the brick invert as required.

Then came my first experience of underpinning. We drove the long access passage under the supported track to the older concourses and then broke out on either side in alternate headings so that bricklayers could pin up the joists which carried the track and the platform walls off the remains of the invert. The whole passage was lined at sides and bottom with flat flanged cast iron plates, caulked with lead to hold out the water when pumping stopped. These merged into special castings which evolved into cast iron rings which duly bolted up with tunnel rings.

The next step was to drive two 12 feet diameter escalator tunnels with a ten feet six inch stairway tunnel between them to connect to the new station tunnel. This had been driven in compressed air though only the top few feet was above the London Clay. On these tunnels I was again involved in a *first* as Dalrymple Hay decided to attempt to solidify the ground above the escalator tunnels by the François Cementation Process. This had unforeseen influences on my future progress but comes in a later chapter. In later years I mentioned at a meeting of the I.C.E. that when one of François pressure hoses burst I was baptised into the injection process world by total immersion.

My first experience with the Press occurred when a reporter asked if our tunnels had joined up and said that he did not need to go into a tunnel to write about one. I told him nothing. The station tunnel was empty when we reached it and we merely checked our position. The journalist wrote among other things "A mighty cheer went up and hand grasped hand through the hole". A useful lesson.

The François company drove hollow drill rods by means of jack-hammers outside along the arc of the short sloping tunnels and injected a mixture of cement with increasing volumes of water at high pressure as the rods were withdrawn.

When tunnel driving started under this protective arch, exposure showed that hard boulders of concrete had been formed, but with areas of untreated sand between them. Conditions were thus worse than if the whole soffit had been untreated, as the hard lumps were hard to trim and sometimes dropped out, exposing running sand above them. So careful timbering was needed.

This was an early proof of what Dr Terzaghi pointed out in a conference at Harvard in 1936 — that cement grout cannot be injected into sands the *10 percent* size of which is less than 0.5 millimetres for a loose state of packing or about 1.4 millimetres for dense sands.

Within a few weeks of starting on Charing Cross, early in 1926 my firm started a three year contract to rebuild Piccadilly Circus Station. For some months we carried on both simultaneously with a staff consisting of J.H. Segrave as Agent and myself for setting out, and no one else!

The mention of underpinning has produced a mental flash-back. I have already quoted examples from my youthful quarry of information, the three volumes of Engineering Wonders which were written by experts for adult readers. One section is headed *Great Underpinning Achievements* by W.T. Perkins. One example is the long-forgotten underpinning of Winchester Cathedral by diver. I was particularly interested as I was living, as a boy, in Winchester, while it was going on, from about 1908 till 1911. The diver would emerge to talk to the public to raise money as a tourist attraction, and by selling polished pieces of perfectly preserved beech-wood.

The Cathedral had been built, like others, on rafts of beech wood logs which were laid on soft ground directly ground-water was encountered. A diagram showed the work to be done by the diver underwater — to cut a series of pits through six feet of marly clay and eight feet of peat down to the gravel. After tunnelling under the walls he laid concrete in bags under water. When this had set the water was pumped down enough to pin up with a few courses of brickwork. In this way I became subconscious of ground-water and the softness of earth at an early stage. In those days Sir Francis Fox succeeded by not lowering the ground-water which would have hastened settlements. Today St Paul's Cathedral with equally dubious foundations on Fullers Earth some feet above the gravels is protected by regulations controlling local ground-water levels, curiously enough on the advice of Freeman Fox & Partners.

Another case was that of the Church of St Mary Woolnoth. In 1924 I spent two short but productive weeks relieving our man who was installing escalators at the Bank Station, on the Central Line. I found that the work included alterations to the old City & South London Station which was under the Church of St Mary Woolnoth. This reminded me so I re-read the detailed description of this remarkable work which made the construction of the station possible in an area where no other land was available. **Figure 32a**.

The article said "A bold and singularly competent group of men were the engineers — the late Sir Benjamin Baker, Mr David Hay and Mr Basil Mott and

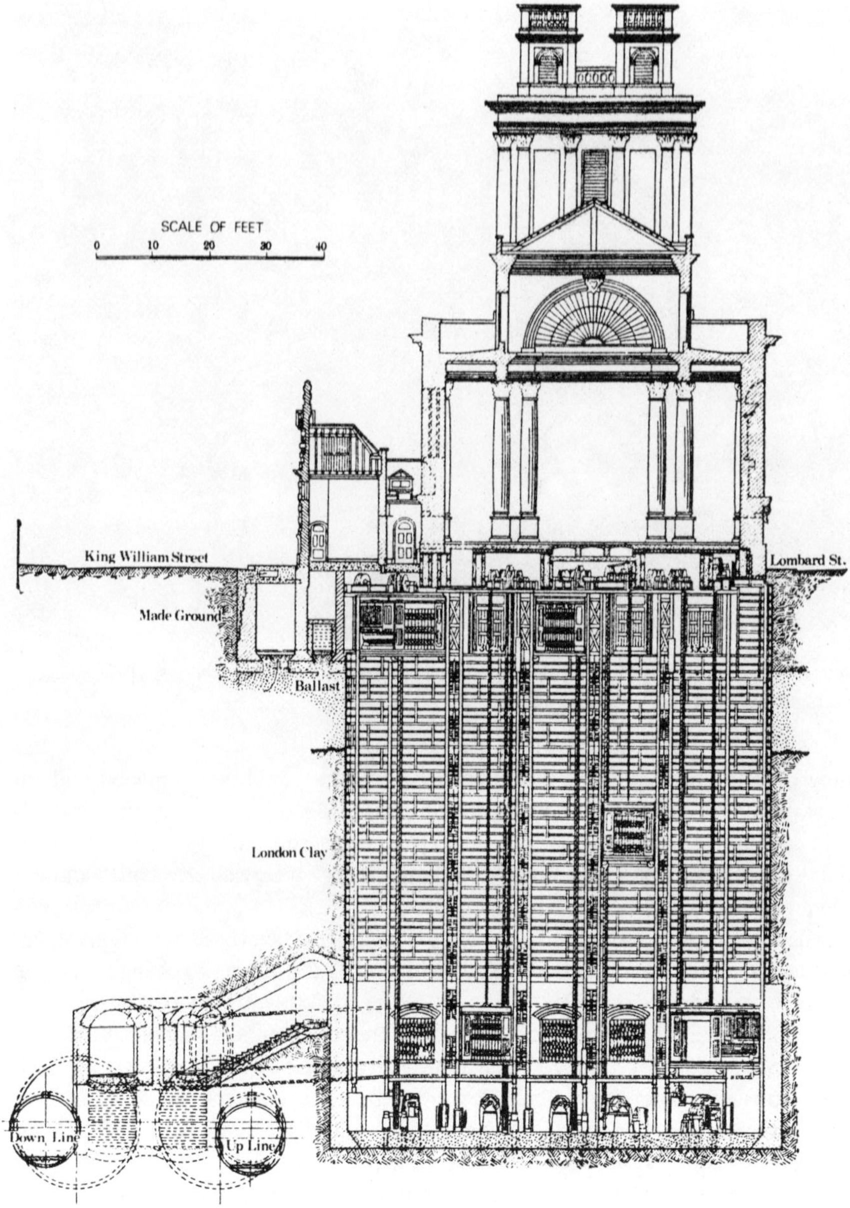

FIGURE 32a. *City & South London Railway — Lift shafts at Bank Station after underpinning St. Mary Woolnoth Church, 1898. (Mott, Hay & Anderson)*

FIGURE 32b. *Site of Bank Station Central Line, 1898, before start of Booking Hall concourse by cut-and-cover methods.*

they gave the most positive assurance that the task could be completed without imperilling the fabric in the smallest degree. They were as good as their word". The description was so good that, having seen the site for myself, in a burst of naive enthusiasm I began to describe it to my Chief, Mr Rowell. He said "I know, I did it myself".

The main Bank Station, with its vast concourse, had been built in sections by cut and cover under the wide open space in front of the Royal Exchange. A wide excavation was needed under the centre of the roofed-over concourse before the first of the large rings, which accommodate three escalators, could be buried. The engineers, Messrs Mott, Hay & Anderson, had revived a method of support which had also been used on part of the Victoria Embankment and this saved the need to strut across a twenty five foot gap. First, a line of five feet diameter cast iron lined shafts were sunk on either side of the cutting. Then as the dumpling was dug segments from a larger type of lining were placed horizontally so that they formed shallow arches springing off the outside of the shafts. These were so rigid that no cross struts were needed, giving an open area in which to work. **Figure 32b**.

Mott, Hay & Anderson again used such a method but with horizontal 6 foot tunnels with a roof of segments for a tricky piece of work at Mile End Station, and Stratford. (Groves, 1946).

Talking of small shafts, two other little jobs come to mind at the same time. Anglo-Persian Oil Co. were building Britannic House at Moorgate so my firm sank a number of small diameter shafts inside the existing basement so that they could be used to carry the load of the heavy steel columns below the tube running tunnels.

We were doing the same inside the basement of Peter Robinson's old Nash building in Oxford Circus, before they rebuilt it, in order to carry loads below the running tunnels. Primitive plant was still in use as winding was done by Jack-roll. This consists of a wooden barrel carried between timber A-frames and rotated by hand with crank handles at either end. This reminded one of the old song *Pussy's down the well*. For a heavy-weight miner asked our light-weight chainman to lower him down, to save using the ladder, when holding on to the rope with one foot in the bucket. The chainman started to lower him away with his own back against a wall. The weight took charge and he lost control of the handle which hit him under the chin. This knocked him back against the wall. He rebounded in time to be hit by the next revolution of the handle. This rhythm continued until the miner reached the bottom at speed while the chainman took the count of ten. He revived with little damage. Factory Inspectors were less visible in those days.

In recent years the courage to spend money has led to the development of large diameter bored piles by different methods and plant. They can make the sinking of small shafts by hand obsolete except in restricted headroom but there is still something to be learned or adapted from our predecessors.

I have emphasised how my early reading, before starting to earn my living, had the happy result of unexpectedly briefing me before I started work on several different projects. These three volumes have today been a help in studying the details of Tunnelling History and are a cherished possession. On looking back, it seems as if they were of more practical help to me than some abstract and slightly incomprehensible lectures of one's Professors.

PICCADILLY CIRCUS STATION (1926–1929)

The reconstruction of Piccadilly Circus Underground Station provided the first, and possibly the only, occasion when a large underground Booking Hall and circulating areas had been dug out and roofed over entirely by tunnelling methods without disturbing the road surface. The roadway consisted of unreinforced concrete and in places was found to be barely nine inches thick. Luckily this was before the days of juggernauts, and horses hooves were clearly audible to us below. Other station booking halls were built by occupying sections of the road and using *cut-and-cover* methods (Hall H. 1928/29).

In recent years new expedients have simplified methods and provided other solutions. At Oxford Street and later at Bond Street Stations series of bored piles were quickly concreted at night with little effect on traffic. Then the roads were closed over a Bank Holiday weekend, prefabricated welded steel units were brought in and quickly bolted together, resting on the bored piles. This provided a flat dome over which the traffic could flow.

When I was giving a lecture in Paris to the Société des Ingénieurs de France, I showed the Transport Board's film of this work. When the first omnibus rolled over the dome the French audience burst into spontaneous applause.

This dome made it possible for the booking hall to be dug out as if in the open and the permanent steel roof was simple to erect and concrete in place. Then the dome could be removed in another single week-end. This is a contrast to the work which is now to be described.

The Artist's impression **Figure 33a** gives some idea of the work which, from its position was much in the public eye. On it the genius of Dalrymple Hay was fully extended, sometimes for the sheer pleasure of extreme ingenuity. His great virtue was, not only designing but in laying down exactly the method to be used.

The picture does not show the complex of twelve foot diameter pipe subway tunnels which totalled one-third of a mile and which twisted round the booking hall, driven sometimes uphill in gravel, as well as the sewer diversions in timbered headings. The four escalator tunnels at two levels and their chambers were built by increasing the size of pilot tunnels. These started as *novelties* — they are square headings with bulging sides as they are made of four segments from a six segment ring, with joints of tapered wooden blocks through which long bent bolts are used. The segments then live to fight another day at their full size. Some pilots enlarged from these were 14 feet in diameter while the escalator tunnels themselves were 22 feet 4 inches, and the chambers 25 feet or 26 feet diameter. Innumerable tunnels of this kind have since been built but these were early days.

Before the days of rotary traffic there were two islands in the Circus. Eros and his fountain were banished to a London Park and we were given his island as our working site. The historic Flower Girls were demoted to the other island which covered the public conveniences. All the work was done through the 18 feet future stair shaft which was half under the road and half in our site. My task was to design the layout to accommodate crane gantries, muck stages, storage space for many sizes of lining and the other usual buildings. We placed three rough office huts on the roof of the mess room for engineers, foremen and Inspectors. **Figure 33b**. This was before the smart portacabin and other portables had appeared.

Much ingenuity went into the sewer diversions, with junction chambers to live sewers, using old rails as bars, and steeply sloping drives for building tumbling bays. When the final junction with the live sewer was being made a miner took out one brick too many and found himself peering into the Bar of the Criterion Theatre.

The pipe interceptors to divert the services into the pipe subway were a novelty. They were over fifty feet long, about 10 feet wide and 8 feet high inside. The lower

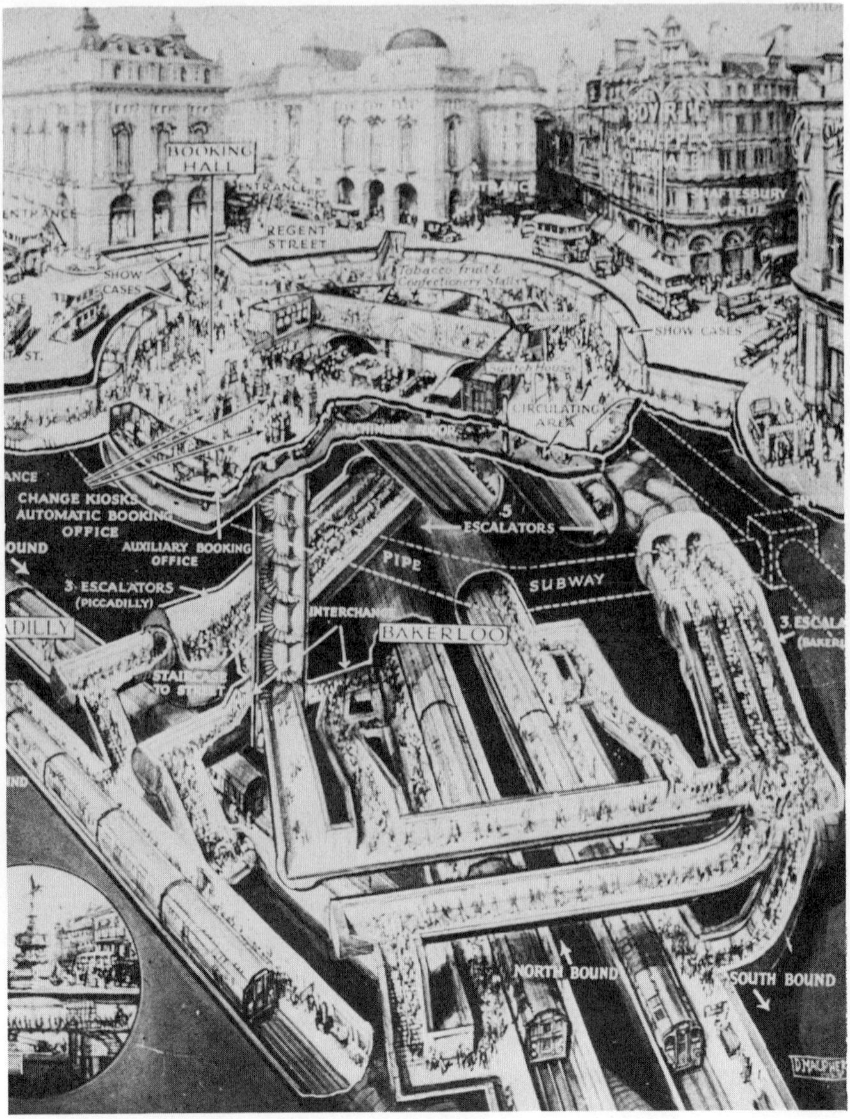

FIGURE 33a. *Piccadilly Circus Station — Artist's impression.*

FIGURE 33b. *Piccadilly Circus Station, 1926 — Working Site.*

part of the walls were built as usual by bricklayers working backwards in 5 feet 6 inch high timber headings. The miners then took out short lengths using 14 feet long joists at close intervals to support the road concrete, which were being held by props until the bricklayers underpinned each length. Any pipes or cables were bricked into the walls until the time came to divert them into the pipe subway. Brickwork is ideal for this sort of work as instant support is given with no need to make shutters and wait for concrete to set.

As so much timber had to be left in to hold up the outside of these many headings Dalrymple Hay introduced steel polings, ¼ inch thick to be used on the outsides and left in, rather than the normal 1½ inch timber board.

The same method of heading-driving and building walls followed by working to complete the upper portion was used for each of the five passenger access subways and the walls of the large new mens lavatories, later to become notorious as a haunt of *junkies*.

The booking hall curved walls were built in three lifts of brickwork, two of them in headings driven one above the other, from several different small shafts sunk inside the booking hall area. All these headings involved nearly 3,000 feet of driving from over 30 different faces. **Figure 34**.

The whole of the steel work for the roof of the Booking Hall was trial erected by Joseph Westwood & Co. in their yard in the Isle of Dogs on the spot where I.K.

FIGURE 34. *Driving timbered heading in gravel — Piccadilly Circus.*

Brunel launched his Great Eastern steamship. Each piece was numbered and called up for delivery at night as required. The main members were plate girders 4 feet deep and 25 feet long. They were dragged underground and placed by means of rollers, jacks, block and tackle and brute force in confined space.

The space was produced by driving a series of five feet wide headings, with pitch pine side trees 6 inches square and capped by 6 inch × 3 inch steel joists which eventually formed part of the roof. The side trees were capped by Yankee Brobs whose Z shape served the purpose of stretchers. Pits for the main columns were sunk in them down to grillages, sometimes previously set in small caverns driven from the pipe subway or the headings for the walls. The centre portion consisted of two squares of 25 feet long main girders sitting upon six 20 feet columns over the deeper section of the machinery chamber. The space between them was filled with slightly smaller plate girders at 5 feet intervals.

As each girder was fixed it carried the 6 inch × 3 inch joists which in turn support the road concrete by packings. After the first main girder was dragged into place the four lesser girders were carefully erected and bolted to it. As they then carried the road concrete they had to be set precisely as the other main girder had to bolt up to them when in turn it was erected.

Eventually we erected 400 separate steel girders, large and small, each in its own piece of tunnelling so we will not follow each one. The outer fringe of lighter beams were supported on timber packings while carrying the road, while bricklayers pinned them up with the last lift of the brick walls.

The Artist's impression has a curious story. Soon after starting the work a friend and I spent six months making a model of the future station. Half way through we became engaged to be married. The model was shown at the Institution of Civil Engineers Conversazione and was a centre of interest. Dalrymple Hay was so pleased with it that he rang up Frank Pick, the Underground Chief, and told him that he ought to buy it. I was summoned by Pick who kept me kicking heels for two hours and then rudely said that he did not want my model. This was infuriating as I had not asked him for the interview.

We loaned the model to the Science Museum. Then the Curator rang me up to say that he had found two of Pick's minions surreptitiously trying to photograph the model to help in an artist's impression, without the permission of the Curator or myself. He sent them packing. I then made them pay me at least enough to cover the cost of the model including the mahogany and glass case. There is internal evidence that the model was the basis of this much published worms eye view. Years after, I got much pleasure from the reported comment of Sir Winston Churchill on Mr. Pick — "Never let that impeccable Busman into my sight again".

One unexpected event was encountering a dry bricklined 7 feet diameter well just outside the London Pavilion. Archaeologists dated the suction pipe as about 1750. The top was a brick dome with a hole closed by a slab of Portland stone. The curious fact is that this was only less than three feet below the road surface but had never been found when all the cables and pipes were being laid over the years on either side of it.

One of the live sewers which had to be diverted ran under a half-round brick lined pipe subway. These had been built all the way up Shaftesbury Avenue when this rather mean street was being carved out of the Soho slums at the same time that Haussmann was building the famous Paris Boulevardes. The diversion had to join another sewer which ran from Piccadilly to Coventry Street in front of the Criterion. Both sewers were egg shaped, 4 feet high and 2 feet 8 inches across under the semi-circular top, and had to be carefully surveyed.

To help us in this unsanitary task the London County Council provided several large red-faced men in heavy donkey jackets and thigh boots who are called flushers. While one flusher crawled around with us in nine inches of dry-weather flow the rest took up positions at other manholes up-stream. We were told that if there was a sudden heavy rain the sewer would quickly fill up with rushing water. The other flushers were there to warn us so that we could make a quick crawl to the nearest up-stream manhole!

On the hottest afternoon of 1926, John Leigh Hunt, the Assistant Resident Engineer, and I, spent three hours levelling inside the 4 feet × 2 feet 8 inch sewer alongside the Criterion Restaurant. The cooking water which kept emerging from the tide-flaps which protected the house connections was much worse than the natural content. The sequel is best left to the cutting from the TIMES thirty years later in December 1957.

TWO MEN IN A TUB IN PICCADILLY CIRCUS

AN AFTERNOON BATH RECALLED

How an eminent civil engineer took a bath in the middle of Piccadilly Circus on a summer afternoon some 30 years ago was recalled at the Institution of Civil Engineers in London yesterday, when Mr. H.J.B. Harding, who was closely concerned with the construction of Piccadilly Circus tube station, described to an audience of young people how he and a colleague spent an afternoon surveying a sewer under the Criterion.

When they emerged into Piccadilly, he said that they were covered with slime and left *abominable snowman* footprints as they walked across to the contractor's workings, situated in the centre of the circus, where they both took a bath in the engineer's shed. "I believe we are the only people who have appeared in the nude in Piccadilly Circus at five o'clock on a summer afternoon", he added.

WORK IN COMPRESSED AIR

Battersea Cable Tunnel (1930)

This 10 feet internal diameter cable tunnel went so smoothly that only a few lessons emerge. The north end started with only a few feet of cover and dived on a sharp curve under the embankment wall, using hooded shield and clay pocketing. The junction with the drive from the south bank, was in London Clay with a precautionary 10 lb pressure, though no buried channel was met. The error was 1/8 inch for line, 3/16 inch for level and 1/4 inch for distance which was my last setting-out before being elevated beyond such matters. This called for triangulation across 1,000 feet of river, and levelling across using a movable target to sight on.

The personal interest was to find that the Resident Engineer on the Bakerloo Tube across the Thames at Waterloo in 1898, who was considered an expert in tunnel survey, had carried his lines and levels all the way to Westminster Bridge and back rather than to triangulate and level across the Thames. I found that from the need to use short backsights and long foresights across the river from one bank, and then repeated from the other a constant apparent error crept in which was due to the curvature of the earth.

On the north bank, due to misplaced economy by my superior, two old reciprocating air compressors were installed for the low pressure supply which delivered directly into the tunnel without passing through air receivers. This caused so much discomfort from pulsations that two rotary compressors which were equipped with their own oil separators and after-coolers which acted as receivers, were used instead at twice the running cost. But the actual cost would have been higher with the old ones as it was becoming impossible to work with them in action.

A derelict shield was found in the adjacent Metropolitan Water Board yard. It transpired that it had been used by my firm many years before to tunnel across the Thames in the London Clay (in free air at sufficient depth) in order to carry water mains. The London Power Company then used it to pass hot water for District Heating from the Battersea Power Station to the new Westminster Council flats on the north bank.

The original paper on the Bakerloo Tube has a longitudinal section, which has often been reproduced, of the drive under the River Thames through a depression in the London Clay filled with gravel. The longitudinal sections along the Waterloo & City Railway, in the paper on that work, also show a number of such depressions of varying size. So two borings were decreed at Battersea as a precaution against a depression or a buried channel.

Two borings were made by a very insignificant little boring firm using shell and auger tools with a small rig and 4 inch lining tubes. The boring contractor hired two Thames Barges and built a rough platform between them to carry the rig. This was my first encounter with the method and the rough and ready way of treating such samples as were retained in wooden egg boxes. No depressions were recorded in mid stream nor met with in the tunnel but a depression could have been easily missed between two such borings. But this provided experience in how to deal with the P.L.A. River Superintendent for permission to moor as he figured a good deal in my later life.

If anyone wants further proof of the eroded surface of the London Clay there are two startling examples of depressions met with in building Battersea Power Station which seem to have still baffled geologists (Berry & Dean 1937).

The Power Station was built in three stages and twenty years after the Cable Tunnel there were two more revealing encounters on the site. The main foundations were carried on Vibro Piles driven through the gravels into the London Clay. After the second half of the Station had been built it was decided to build heavier modern turbo-generators on the original bases. Mr Vernon Bartlett was Chief Engineer of the London Power Company just before nationalisation and, at his request, I drove headings under the concrete bases among the upper part of the Vibro piles. These were all sound with no sign of necking when the long lining tube had been withdrawn.

From these headings we drove down injection pipes and solidified the sands and gravel with chemicals down to the London Clay to increase the bearing capacity of the piles and the base. The headings were then filled with concrete to underpin the turbo-block bases.

The second episode has a message. It occurred while we were providing the foundation for the fourth chimney. Shafts are needed to start tunnels so I regard shafts as vertical tunnels. A thirty feet square shaft was being sunk through the usual Thames sands and gravel in dry ground using three settings of 3 inch timber runners with massive timber walings and struts. I had an assistant who was highly intelligent but over confident.

I instructed him to be sure to lace, hang, and prop every frame as the work went on. As the ground was so good he asked me why, and what could possibly happen to make this necessary. I said that I had no idea but he must take these precautions against the unknown. When the excavation was nearing the bottom I visited the site and found a serious collapse being narrowly averted.

The walings had not been laced, hung or propped, as he had relied on the friction of the wedges to hold the timbers. One frame on the east side had slipped down nearly two feet and a crane was hanging on to the struts to save the situation.

The cause is full of admonition. The runners had nearly completed their work, driven down in beautiful ground conditions when suddenly soft puddle clay started to spread out from beneath their toes. Unknown and unseen to us a trench of puddle clay had been sunk years before only two feet behind our timber face.

I knew that the original site had been surrounded by steel sheet piling so that the ground-water could be removed but what I did not know was revealed on referring back to the 1935 paper, which read — "Concrete retaining walls were founded in London Clay built inside the steel sheet piling all round the building — with the exception of the east side of the Boiler House where a puddle wall was constructed in trench down to the London Clay to act as a temporary seal until the second half of the Station was built" (Berry & Dean 1937).

We were part of the second half of the building but there was no sign of this trench on the Power Company's contract drawing for the chimney foundation. Twenty years had elapsed and nationalisation had changed the personnel. My instinct and training under sound engineers had led me to provide against the unknown, even if partly frustrated by a young man who had the lesson well rubbed into him. Perhaps we should all read Institution Papers with more zeal where they are appropriate.

FORD'S POWER HOUSE, DAGENHAM (1930–1932)

Henry Ford arrived in person to choose contractors to build his new factory beside the Thames on the abandoned Corporation of London rubbish dump at Dagenham. This was to be completely self-supporting as it included its own blast furnace, fed with iron ore, which was to be discharged at their new 1,500 feet long jetty. The key to the scheme was its own Power House of 60,000 kw which was later increased.

Most modern power stations are carefully sited in as favourable conditions as the site allows. In 1930 someone in Detroit drew a rectangle in the only space left on the map after setting out the main buildings and wrote *Power House*. This left the Engineer (Sir Cyril Kirkpatrick & Partners) no choice. It was outside the river bank, on forty feet of soft estuarine clays and peat, known locally as bungum with plenty of water-filled sands and gravel below. Eight boreholes recorded this sequence. The tidal range which affected part of the site was from +17.30 OD to −11.66 OD (Allin & Nachshen 1932).

The turbine house, 305 feet long and 86 feet wide, was to be built in a cofferdam of No. 3 Larssen piles driven to −40. OD which was divided into a number of compartments. The east end adjoined a narrow earth river bank which protected a vast area of ground behind, below river level. A tidal sluice allowed the small River Beam to discharge into the Thames at low tide. As this area was vulnerable the Engineer decided to avoid heavy pumping (1930 being before the days of ground-water lowering) so the specification for the Power House included a sub-contract already negotiated, for the solidification of the whole area below formation level inside the cofferdam by the François Cementation Process. This was intended to withstand 30 feet head of water.

The actual organisation and execution of the injection work was exemplary. The specified thickness varied from six to eight feet over an area of 2,285 square yards.

In the work-shed, 6 cementation pumps, driven by steam or compressed air were served by two boilers and two diesel air compressors. Sixteen mixer tanks fed the pumps and 6 pairs of drillers using jack-hammers drilled 3,300 holes at 2 feet 4 inches centres to a depth of about 44 feet.

The pressure used for injecting was 300 to 350 lb per square inch, but lowered later to 150 or 200 lb. 3,400,000 gallons of water was supplied to mix with 1,405 tons of portland cement, about 15/1 by volume. Considerable quantities of wasted cement were later found accumulated on the foreshore.

Unhappily, when the first compartment digging had begun to expose the top of the treated gravels and sand a powerful *blow* occurred where areas of sand had declined to accept the cement, which had preferred the coarse material!

At this point the whole work was becoming a shambles. The 75 feet long 18 inch octagonal fondu concrete piles which were being driven inland behind the Turbine House to carry the Boiler House were found to be pushing the cofferdam towards the river. A six ton *monkey* broke from the top of a 90 foot high pile frame and disappeared 25 feet into the *bungum*. A thousand tons of aggregate, brought in by Thames sailing barges from Brightlingsea, had been dumped on the ground near the river's edge. Not surprisingly it set off a classical Swedish slip curve, destroying the timber piled unloading crane jetty. In trying to excavate the compartments, two draglines had unwisely been used on the soft ground instead of loco cranes on track. These had trodden themselves almost out of sight. As they sank deeper more timber planks were fed beneath them to a final thickness of nearly ten feet.

At this juncture, at the early age of 30 I was sent, at a moment's notice to take over as Contractor's Agent. For a few months we struggled to put matters right.

Then, as we were bottoming-up the second compartment a similar *blow* occurred which was so much more severe than the first, that the hole was allowed to flood. The bottom was concreted under water by bottom-dumping skips while divers struck the timber frames as the fondu concrete level rose. The heat of the setting cement produced an interesting volume of water vapour.

This blow produced a considerable atmosphere of misery and gloom as, not only was the Power House a vital factor in the working of the factory, but also there was the risk of breaching the narrow Thames river bank by the River Beam with the possibility of flooding an appreciable area of Essex. This caused some panic.

There is a small inland lake behind the factory called Dagenham Breach which was formed after Captain John Perry R.N. had closed the breach in the Thames Bank which had produced it, where Samuel Williams Dagenham Dock now stands to the west of the factory. But on reading Perry's account of his work published in 1721 one of the Engineer's staff came across the fact that the famous Dutchman Vermuyden, in the interval of draining the Fens, had closed a previous breach in the Thames in 1600.

This turned out to be exactly and precisely where we were trying to build the Power House. The boreholes had missed it but later we dug into some of the timber boxes filled with chalk which Vermuyden had sunk to close the breach — a forecast of the British return by supplying surplus caissons from Mulberry Harbour to help to close the Dutch Dykes in the floods of 1953. So on Henry Ford's own doorstep we were able to prove that History is not bunk.

As the designers intentions had been frustrated the whole method of working was changed and the work repriced by agreement. The critical decision was taken to complete the work below ground under compressed air. The Firm's accumulation of such experience had just been enlarged by the advent of Sir Henry Japp K.B.E., who had made a name by driving the East River tunnels in New York for Lord Cowdray's firm of S. Pearson & Sons.

This allowed us the chance to obtain ample plant and to redesign the whole lay-out of the working site, which is the key to any successful contract. So what had looked like a period of work in the open air turned into another form of unusual tunnel work.

As the cementation work, for all the trouble taken, had proved a failure (but not to my surprise after Charing Cross) the whole of the base of the turbine house was completed under an air tight floor. Fifty six cylinders, 8 feet 6 inches in diameter were sunk between the timber frames, six in each compartment, with other variations, each with its own simple vertical air lock. The bottom of the cylinder was concreted to a safe level which left sufficient clay to hold back the head of water. It was found that this only varied within a foot or two of +2.00 OD in spite of the rise and fall of the tide. A reinforced concrete floor was then laid on these supports to form an air-deck and connected to the steel piles by short lengths of welded joists. Air shafts were concreted into the deck and the space beneath dug out under compressed air into the gravel, concreted and the space between this and the air-deck grouted up.

Compressed air caissons were used for the pump well, screen chamber, intake and outlet shafts for the cooling water. Twelve ton electric Scotch Derricks with 120 feet long jibs were installed which covered the site and quickly moved air locks, and other plant from place to place. This made me a strong advocate of such useful, strong, easily erected plant and with a wide radius of action; as well as a strong attraction to air caissons as an expedient in proper conditions.

The experience on this site was naturally intensive and the whole affair had subsequent after-effects little expected at the time. There was one case of light relief. Sir Cyril Kirkpatrick was at the time President of the Institution of Civil Engineers so had much on his mind. He was also a serious character and a Churchwarden. At a site meeting, Mr Boyd, the Chief Engineer for Fords on the work was talking about how the whole thing had changed, Sir Cyril replied "We did not know that's what would happen". To which Mr Boyd riposted "Ha! that's what the girl told the Sailor". Sir Cyril blushed, to the delight of us younger men in the presence of the Great.

The conditions on the river bank at Dagenham were rigorous. The adjacent City of London rubbish dump was still in active use, reaching heights of over 100 feet. The plague of flies made it necessary for us to build a kitchen for staff meals, heavily protected by wire mesh. In winter it was chilly and wind swept. In those days, except for some key operatives, no protective clothing was provided. Engineers had to buy their own and organise and pay for their own meals and transport.

General Caisson Work

Figures 35a and 35b show the general layout and the two large steel caissons for the pump well and screen chamber, as well as the 14 feet diameter intake and discharge shafts. These were sublet to Dorman Long & Co. whose staff and plant had just been released from the caissons for Lambeth Bridge. Observing these and the adjoining skip-hoist pit caisson made me an enthusiast for compressed air caissons.

The working chambers were eight feet high and, above them the steel skins were supported by light structural trusses. Concrete walls were built inside these which provided added weight as the caisson sank. The pump house caisson was suspended all the way on lowering gear as eighty feet long Larssen piles had to be burnt off under the cutting edge.

The main surprise was to find that caissons only require about one-tenth of the air needed for a vertical face of similar area. Also as the pressure increases with depth the work at the highest pressure is only for a few shifts at the end of the sinking. The friction is reduced by the passage of air around the skin, compared with that found with monoliths, though this has been improved by use of bentonite injections.

The pile driving had caused dispute as a number of piles broke and refused at two thirds of their drive. Then the floor of the screen chamber caisson reached that level and revealed a bed of naturally cemented Thames gravels (caused by ferrous salts) known locally as Blackwall Rock. This is apt to be broken up by the method used in old-style boreholes and so is not revealed by them.

A tragic lesson was learned in the sinking of Dorman Long's Intake Shaft. This had sunk in free air to the depth at which the air deck was to be fitted. Compressed air from the compartments was bubbling up through the soft clay. Dorman Long's engineer was sleeping at nights in his office hut. That evening, in my office and presence, he strictly forebade his Irish ganger to go near the shaft till next day.

An air pump had been holding down water which was entering. This pump stopped in the night. The Irish ganger went down, in spite of instructions, to start it up. He collapsed into a foot or two of water. One after another five members of the Irish gang went down to the rescue and collapsed in turn. Mowlem's night-foreman saw something was up and was in time to stop the sixth from following. Tests proved that there was no gas, it was entirely the lack of oxygen which caused the collapses. The peat had absorbed the oxygen as the air passed through the ground. The bodies were recovered by a diver who was working on the jetty contract.

As a result, all sorts of precautions were rained down from Head Office varying from canaries to Davey Lamps. The point was missed that there was no gas and no lack of oxygen inside the workings under air pressure. White mice were obtained for use inside the compartments. On entering one, the ganger told me that his

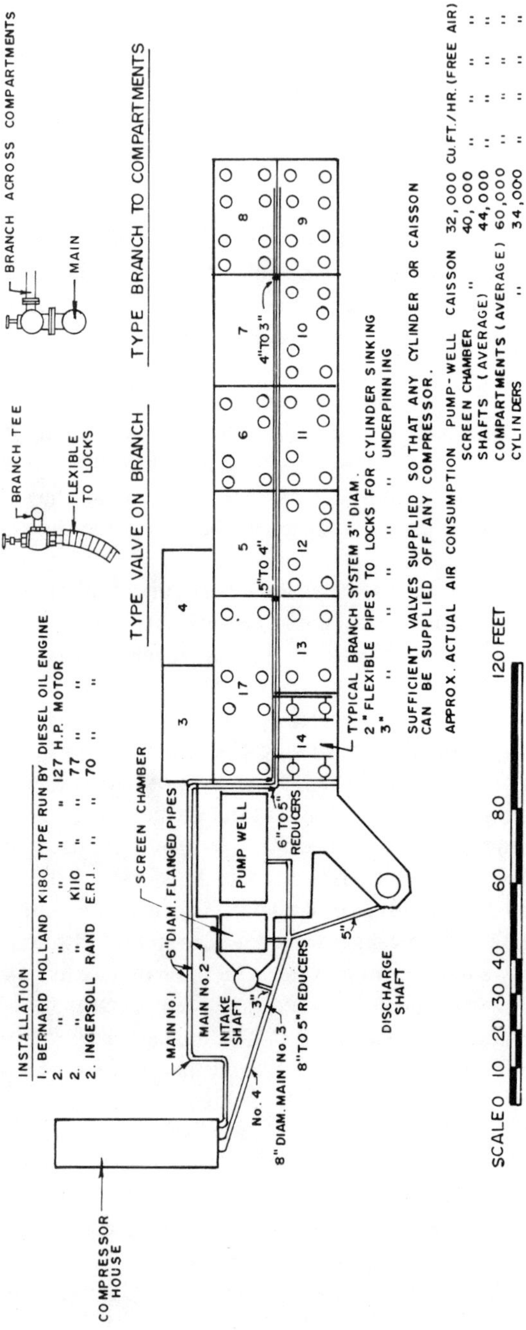

FIGURE 35a. *Ford's Power House, Dagenham — Plan of coffer dams and caissons.*

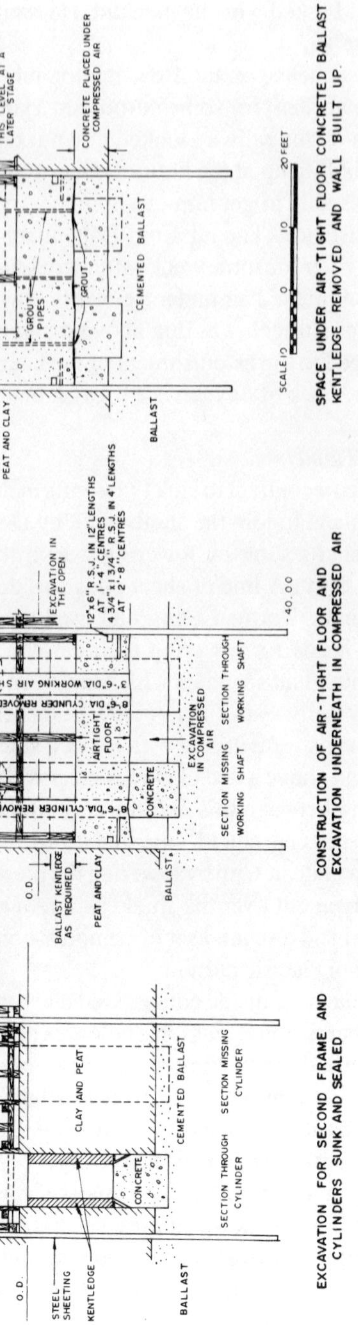

FIGURE 35b. *Ford's Power House — Construction of chambers in Turbine Basement.*

mouse had died. I asked what he then did. He replied that he sent to the office for a new one. **Figure 36**.

There was a sequence on the Bow-Leyton tube contract. The Transport Board had sunk their own shaft for some purpose in a site near the tunnels. One weekend two plate layers on the railway looked down it out of curiosity and saw air, from our tunnels, bubbling up at the bottom. One went down to see and collapsed. His mate luckily was able to get him out in time. Air can travel under an impervious bed for some distance. The air from Dagenham bubbled up at Erith, two miles away across the river for some weeks after work had stopped. In Jersey a sewer was driven under compressed air under a cover of clay described locally as bulls liver. A few years later a trench was dug above it and compressed air, which had been bottled up escaped to the astonishment of the later contractor who was unaware of the earlier work. Lack of oxygen is a big risk in abandoned tunnels.

Cooling Water Tunnels

We were awarded a contract to sink two shafts in the Thames and drive two cooling water tunnels inland to join the shafts sunk by Dorman Long. These were 14 feet diameter steel shafts sunk by lowering gear from a timber jetty built alongside Ford's concrete jetty. A line of sheet piles was driven to separate the intake from the discharge shaft. Dorman Long had used their 100 ton hydraulic jacks with safety screwed heads for lowering their caisson. We were equipped with large screwed rods with giant spanners which had been preserved from the shafts on the Waterloo & City Railway in 1893—forty years before. They served their purpose.

The 7 feet tunnels were driven with hooded shields. The top half of the face was in gravel and sand above a bed of clay. The gravel was so open that water cascaded down the parting between this and the clay, though the air sustained the closely timbered face. As one shield emerged from the opening in the shaft a blow occurred. This was dealt with by lowering the pressure so that the ground collapsed on to boards driven out over the shield skin. An air main led out from the central compressor shed and a second set of compressors was installed on the jetty with a different source of electric current.

Fire is a constant risk inside compressed air work. In order to connect the screen chamber to the pump house caisson a hole was being burned inside it. Presently the temperature rose and, on looking round, a belt of red hot steel could be seen round the caisson. The burning gear had set fire to a thick bed of peat, fed by escaping air. Luckily it was possible to take off the pressure and let the ground water rise to put out the fire before it spread under the factory.

But a lesson on a fire outside the caissons was also learned. As a precaution against fire I had built the new compressor shed out of tubular scaffolding clad with galvanized sheeting which also gave 100% salvage. The extra electric supply was connected to the same distribution board as the existing one in the electricians

FIGURE 36. *Ford's Power House — Work in compressed air around cylinders under floor showing air shaft (and white mice).*

wooden hut which Higher Authority would not allow me to rebuild. One night of fog an electrician on duty as a safeguard, upset a can of oil and set fire to the hut. This cut off all electric supply to the land work. The tunnels luckily had their second supply. The diesel compressor was able to maintain the air under the decks so no untoward events took place.

The chief interest is the matter of the fire fighting. Hand extinguishers were in the electrician's hut but he could not reach them. A 3 inch water main fed the adjacent concrete mixers but it ran under the burning hut and only steam came out. A pump was in an adjacent water-filled cofferdam but it was electric. A long-jib derrick, for unloading aggregate from barges, also could cover the fire but this too was electric. There was a steam crane on a timber gantry for feeding the hoppers but the driver had to back out of reach as the fire had spread to his timber gantry. The only approach to the site was by water or rail, except on foot for half a mile in front of the factory. By the time the local fire brigade had trudged to the site, carrying what portable equipment was available, the fire had subsided.

There seems to be a moral in this story. At least it emphasises the need to have separate supplies of electricity but also diesel stand-by plant. The unexpected does happen.

This Power Station is small compared with modern stations but much else was learned by all concerned on the work other than tunnels, on land and water. Three of the few young engineers involved became Presidents of different Learned Institutions.

The adjacent jetty was built of concrete piers formed inside dozens of 14 feet diameter steel cylinders mostly sunk in free air, with cast iron 2 ton kentledge blocks balanced on the top to overcome friction. In the worst case this reached 500 tons. In one cylinder a diver was sent down to clean the bottom before concreting under water. The tide went out more quickly than was noticed so that an adverse head was thus built up in the cylinder. This blew the diver out under the cutting edge, where he was under more than thirty feet of river mud. His linesman was able to pay out his line and so keep him alive.

Luckily there were two other divers on the work. One diver got in a boat alongside with a water pump and jetted himself down 30 feet through the mud until he grasped the hand of the buried diver. The other diver, inside the cylinder paid out the air line. The line was just long enough to allow the victim to emerge. By then the tide was so high that he had to be held upside down to unscrew his helmet. He was then put in our medical air-lock for a spell but was none the worse. The others were awarded gallantry medals.

Chapter 7

GEOTECHNICAL PROCESSES AND PROBLEMS

THE START OF "PROCESS WORK"

While the battle to overcome the failure of the cementation process to solidify the sands and gravel was being fought, John Mowlem with Edmund Nuttall Sons & Co. were building the King George V graving dock at Southampton in order to receive R.M.S. Queen Mary which was being built, fitfully, in Glasgow. Mr Wentworth Shields, who was Docks Engineer at Southampton designed the Dock.

Being a wise and experienced man, he carried out his site investigation to unusual depths. The excavation for the dock was to be 100 feet deep but he pressed on and found, at a much greater depth, beds of Bracklesham sands containing water under an artesian head which would reach above ground level. This head was capable of making his new work into a floating dock if he had not shown such foresight.

Siemens Bau-union had recently developed the system of Grund Wasser Absenkung by lowering the ground-water by means of deep bored filter wells, each equipped with the new Siemens deep well submersible pumps. Mr Wentworth Shields engaged them to install a system of ten wells run from a central control, before the work of digging had got under way. The result was remarkably successful.

Before proceeding, there is another lesson which we learned soon after. The Southern Railway decided to build a train ferry dock at Dover but for some internal psychological reason left Wentworth Shields behind at Southampton. No borings were made as Lower Chalk was expected on the site. But, unfortunately, the dock was sited in the water above an old cliff fall. The team which had been so successful at Southampton were transferred en bloc, Resident Engineer as well, but were brought to a standstill. The I.C.E. paper on Dover Train-Ferry Dock (Ellson 1937) describes how the work was eventually completed by concreting under water. It is fascinating to read in the paper that reliance was placed on the solidity of the Lower Chalk in the early Channel Tunnel workings. But these are on the other side of the Shakespeare Cliff and are far remote from the site!

The contrast between the trouble at Ford's and the success at Southampton led Sir George Burt, the head of John Mowlem & Co. to make an agreement to use Siemens Bau-Union Patents in this country both for the Joosten two-chemical injection process and ground-water lowering by deep and shallow wells. I was told to be their *expert* so, for five years, I was the sole British prophet to test out and preach the new gospel about their proper use to consultants, contractors and government officials.

It was a curious existence and a slightly lonely one to be the only exponent of a new art, previously unknown in this country. Fortunately the two years spent on Ford's Power House had been a time of intensive experience which included steel and concrete piling, cofferdams, varied and massive concrete work, shafts, caissons and underwater tunnels, considerable compressed air working, including much work in tidal waters, mixed with flood, fire, gas and *earthquake*. As a result I was better equipped to brood over our own, but chiefly other peoples, problems than before this curious episode.

The work which follows brought me into contact with top members of the Professions at an unusually early age in order to discuss their problems and the possible application of the appropriate process. I was acting as a professional engineer and not a salesman peddling a patent cure of which he himself had no practical experience.

The first step was to find what it was all about. I collaborated with Siemens engineer from Southampton for a year or so after a visit to Siemenstadt in Berlin in 1932, a year before Hitler came to power. The first work was to demonstrate and test out the action of the Joosten two-fluid process at Monument Station for the Consulting Engineers Mott Hay & Anderson, where my firm were building escalator and other tunnels to make a passenger interchange between the District Railway and the Bank Station on the Northern Line.

Siemens German Foreman arrived with a set of injection pumps, mixing tanks and injection equipment and proceeded to drive pipes ahead to form a sequence of arches of solidified ground before each six rings of a new twelve foot passage was driven in the sand and gravel, which was dry at the time but very loose. I chose two of my younger foremen as pupils and the three of us spent all and every day down below learning the drill in every detail. The German Foreman spoke just enough English but the miners amused themselves teaching him bad language whose meaning he did not appreciate.

Dr Joosten and Siemens Bau-Union had carried out enough full-scale research to arrive at the best proportions to use and the right brand of chemicals. The drill of driving flush jointed pipes, perforated for the last two feet with a closed driving point worked far better than the cementation use of an open ended hollow drill steel. The measuring tanks and the quantities were all based on the metric system so we kept to this as our equipment multiplied.

Quite a lot of other injection work was carried out at Monument. It was entirely successful and visited by consultants and contractors. From 1932 the first few years were during the slump with a big recession but luckily the London Passenger

Transport Board carried on with their programme of enlarging stations. A number of these were involved, not only with the Flood Plain gravel but also the Taplow and Boyne Hill terraces.

Following our successful demonstrations, among other work, we formed arches of solidified ground over ten different escalator tunnels of 22 feet 4 inches in diameter where they passed through sands and gravel. I christened the process Chemical Consolidation but this was before the days of Soil Mechanics when the pundits decided that this was the wrong use of the word.

The main problem was to obtain pipes strong enough to resist the tendency to bend out of line and also to drive them for thirty feet or more in one length. The pipes would be guided along a sloping plank and driven by a pneumatic paving breaker. The pipes were withdrawn at the same intervals for injecting the second chemical as on the drive. Special jacks were needed to withdraw them.

One rather unusual example was described at length in *Engineering* November 17th, 1933, of work at Knightsbridge East Station, to protect the tunnel driving through the loose ground, the lower part of which was below water level and driven in compressed air.

To quote from the article gives an idea of the atmosphere at that time, "... there was some doubt whether the loamy sand could be satisfactorily dealt with. In order to obtain a measure of protection at places where the process might fail thirty-nine 4 inch steel tubes were driven round the outside of the tunnel at 12 inch centres. As these tubes were pitched at 1 foot 6 inches away from the outside of the tunnel they were in the consolidated ground and would have acted as a support for the latter in case of local failure. If on the other hand the process failed over a large area, it would have been possible, by excavating out to the tubes, to have cross-poled between them and thus obtain a safe and stable support for the overlying ground".

This was entirely the idea of Sir Harley Dalrymple Hay who inevitably haunts this sort of work at the time. It was not needed again. The article later on, "It was found that the chemical treatment of the sand strata had produced the equivalent of a soft friable sandstone which was excavated by pneumatic picks, but was not sufficently hard to impede progress ... the air necessary for the tunnelling operation had been actually supplied from one 2,000 cubic feet per minute compressor, while if the work had been carried out under ordinary conditions some 7,500 cubic feet per minute would have been required".

A chamber of 27 feet in diameter had been provided at the top of the escalator to allow for the pipe driving. To make sure that the 4 inch pipes drove straight and true we fixed a 9 inch × 4 inch timber to the exact 30 degree rake which was moved round a curved rib. The pneumatic driving hammer was fixed to a metal sledge which slid on the plank and a bottle jack was fixed behind it to keep up a constant driving pressure. Some tubes were cored out to test their position. To do this I had a long narrow steel trough made with little partitions across, which could be fed down the tube on the end of an inch pipe. Above it rested a ¾ inch pipe down which we blew compressed air. This dislodged the gravel successfully. Much of it collected in the trough as it fell back but the rest whistled out past our ears in an impressive shower of buckshot. In the event all this proved unnecessary but as the top of the escalator was only just below the busy street precautions at least disarmed criticism. It was not repeated.

It was interesting to listen to one of the competitors for the Harding prize in 1978 describing the method used by Mott, Hay & Anderson when faced with a similar problem but in boulder clay at Newcastle while burrowing under the local equivalent of the Nelson Column. There bored piles of the modern kind were driven which interlocked to form the protective arch before driving the escalator tunnel. The piles were big enough to allow a man to climb down to break out boulders.

Other fears were expressed that injecting would not be possible from inside a compressed air tunnel in case the escaping air prevented the chemicals from intermingling. I was able to demonstrate that this was not so on the escalator tunnels under the Post Office, now renamed St Paul's, Station. This led to much further work in other parts of the country.

The whole injection methods have advanced considerably but these are early examples of the start of what was then a complete novelty. The injection pumps were the Worthington type double acting piston pumps using compressed air but with the water cylinders lined to produce higher pressures. One case was a problem when called upon to consolidate the gap between two compressed air caissons at the outfall sewer for Dublin at Pigeon House. The only access was down a three feet diameter air shaft. Fortunately I found in the Depot two Worthington boiler-feed pumps (forty years old) which were small enough to be lowered on end down the air shaft (Harding 1946).

Later in this period on the continent single fluid processes were being developed which needed accelerators and inhibitors to help them to be effective. But in all such work the main problem is how to reach the area to be treated, regardless of which nostrum seems appropriate. Some are chosen to reduce permeability, some to increase strength.

After the war, specialist firms began to appear. In Britain, Soil Mechanics Ltd., a subsidiary of John Mowlem & Co., formed in 1945 and run by Glossop, Golder and Harding, was the first to combine soil mechanics and use of processes and to be the first to develop site investigations in the modern manner. Collaboration with Soletanche, the French firm who developed clay injections and the tube-à-manchette showed how such methods needed experienced staff and special plant. Also they are best handled by independent firms and not as a side-line of a major firm, reluctant to provide capital.

In May 1963, a Symposium was held at the Institution of Civil Engineers under the title of *Grouts and Drilling Muds* (Butterworth 1963), using the word *grout* to cover the field. This is a standard work.

Owing to my background, in my vocabulary *injecting* is the *action* of driving pipes into the ground in order to deliver doses in predetermined areas using one or more of the many modern prescriptions which are now classified as *grouts*.

To a tunnel engineer trained 55 years ago, *grouting* is the *action* of applying the nozzle from a grout pan or later, a pump to a suitable orifice in order to force a grout (generally cement based) to travel on its own to fill voids behind linings, to fill cavities or fissures in brickwork. The distinction will become clearer later.

At the time of the Festival of Britain exhibition new escalators were driven to serve the tubes at Waterloo Station, after sands and gravels had been treated by a double arch of pipes above the escalator tunnels, using the Joosten process, down to the London clay with complete success. This was in contrast to the attempt to use cementation on previous escalators at Waterloo in 1926 which Dalrymple Hay described as a complete failure. "As excavation proceeded downwards only strips or bands of cementation were found, and where there was sand there was no cementation at all". (Jones & Curry 1927).

Many lessons were learned when pioneering the Joosten Process. At first more work arose from rescue parties for those in distress than from pre-designed applications. It was often most difficult to find out the type of soil to be treated owing to faulty or non-existent site investigation.

Samples had to be judged by eye as modern sizing analysis was almost unknown. We quickly proved the ability of the Joosten Process to deal with most of the sands. It gave useful uniform strength yet was trimmable when excavating.

Injection from inside compressed air workings proved successful, the escaping air did not interfere with the reaction of the chemicals. But running water could disturb them before they could combine in the correct proportions.

An insight into human nature: if an arch of solid ground was formed above a tunnel but ended invisibly some inches above the soffit, clients would worry until probing proved the real facts. On the other hand if the chemicals had spread down and made the face solid clients would then give no thought to how thick the arch above might be, which was the real object of the exercise.

Today every college has its lecturers on soil mechanics and students take the Gospel as handed down with little thought of the early work of this engineering revolution.

In such pioneering work Engineers and Foremen had to teach themselves while they accumulated experience. It was only possible to increase the number as the work load developed and to keep them available at short notice.

When incidents were small simple plant and equipment could be used. When work was upon a larger scale other possibilities of using more sophisticated plant with bulk handling of chemicals arose. In later years personal experience was advanced when called in by the Cementation Company over problems which had arisen during the injection work on Blackwall Tunnel Duplication (Kell & Ridley 1967) and subsequently in helping the case for final settlement. **Figure 37**.

FIGURE 37. *Second Blackwall Tunnel — Artists impression of injection work. (Cementation)*

This injection work was on an unprecedented scale. Basically the ground around the 28 feet 2 inch inside diameter tunnel was to be impregnated through an area of about 50 feet in diameter. Two cast iron lined tunnels 7 feet inside diameter had been driven at the top and bottom of the future main tunnel with hooded shields and an unusually high volume of air. Tapped holes had been drilled in the linings of a diameter large enough to allow gland pipes to be fixed through which the injections could be made in free air conditions.

These pilot tunnels had been driven at a fast rate by using very large volumes of compressed air in the sands and gravels. This was thought to have increased the voids as more chemicals were needed than expected.

The large scale of this work and the cramped conditions to be overcome made it financially possible to install a highly sophisticated system on the surface with a control cabin. In this, three control panels were used with electronic and visual aids with recording equipment. A site laboratory monitored the mixing of the chemicals which were pumped from the working site to the several injection points which were in telephonic and electronic connection with the control cabin.

The Cementation Company did not negotiate the Joosten Process but went on to become expert in other injection processes as they developed, mostly single fluid ones. Although the François process proved unsuitable for the sand condition the Cementation process has been the salvation of many a deep shaft-sinking in the South African Goldfields and elsewhere. Later I was able to advise the Cementation Company on contractual problems during and after their most successful work on the second Blackwall Tunnel. But that is another story.

Tunnels for Grouting

These examples of how injection processes have helped in tunnel construction are a reminder matched by the paradox of how tunnels have helped in grouting work. In 1953 Solétanche invited Soil Mechanics Ltd. to tender jointly for the grouted cut-off curtain for the Dokan Dam in North-East Iraq, described as one of the largest ever undertaken. The work included the construction of tunnels for grouting, 3.55 m high and 2.4 m wide, to lessen the cost of drilling and grouting from the surface with other advantages. This exercise produced a monumental volume of remarkable data and an educational exercise, but unfortunately the tender was not successful. The construction of these tunnels and the philosophy behind them and lessons learned can be found in the paper on the Dam, paragraph 129, (Gimson et al 1959). Examples are quoted later of the 7 feet pilot tunnels on the Bow-Leyton tube and on the second Blackwall Tunnel.

Brook Green Sewer

In 1934 a small firm was driving a brick-lined sewer tunnel, 5 feet 6 inches internal diameter in 8 feet 6 inch lengths, taken out *on bars* in London clay. There was a head of 17 feet of water above the crown of the tunnel in the sands and gravels above the clay.

When the face was about to cross under a street, probings showed that the thickness of clay was decreasing rapidly, so cast iron lining was resorted to. The use of compressed air was refused by the Engineer although it was a provisional item in the contract.

When only two feet thick, the clay broke, sand and gravel ran into the heading and heavy pumping was needed to prevent flooding.

On being called in to the rescue the first attempt was to solidify the ground using the Joosten process by pipes driven from the face. But the water appeared to be washing out so much of the chemicals that successful reaction seemed dubious. A timbered heading was driven just above ground-water level so that without breaking the road surface, shorter vertical pipes could be driven. The heading revealed a typical spherical cavity which exposed the footings of a house. This was quickly filled and grouted.

As the opposite face advanced and the clay again thinned out, treatment was carried out by pipes driven through holes augered in the clay face. Then a second heading was driven and the rest of the length through the depression was treated by vertical pipes. **Figure 38**.

During the driving of the heading before the injection work, the tunnel was allowed to flood to still the ground-water. This led to a sequel. At the far end, an 8 feet 6 inch length had been excavated on timber bars ready for the bricklayers, with the customary bar-heading in advance. Here the clay cover was 5 feet thick with 5 feet of waterlogged gravels above the clay.

Directly the tunnel was pumped out, swelling pressures became heavy. The bar props punched over two inches into the horizontal cill. The bars bent and so many props were hastily erected that entry became almost impossible. Then the clay collapsed between the bar props and the start of the heading. The gravel ran in and formed an irregular cavity under the street. This was refilled with granular material. Injections from the street solidified the fall and the filling and then work went on in cast iron lining. This comes under Human Behavior. It was stupid not to allow air to be put on. It would have been cheaper and quicker with no damage.

One of the difficulties in launching the Joosten Process was the need to share profits with Siemens after paying a high royalty to Dr Joosten on the tons of chemicals used. This made the cost per cubic yard of ground treated seem high in the eyes of potential clients. When we were travelling by train to organise some treatment to the entrance of the Southampton Graving Dock, Sir Henry Japp started teasing Dr. Joosten, whose nationality was Dutch, about his royalties and quoted George Canning, "In matters of commerce the fault of the Dutch — is offering too little and asking too much". To this Dr Joosten smartly replied, "I always thought that that was the Scots, Sir Henry".

By good luck and not by foresight our agreement with Siemens Bau-Union ended in July 1939, just in time before the War. A few years later we were able to negotiate the rights of the Guttman Process from Dr Guttman who had managed to escape from Germany and who had already patented his process over here. This introduced a third chemical which increased strength and permeability.

GROUTING

Grouting with cement or lime with water was pioneered by Greathead who patented his grout pan in 1886. The introduction of the François Cementation Process led to the development of grout pumps for this purpose. The problem was to develop a pump which could deliver the cement and water mixture without the clogging up of the valves. The most common form of pump has large ball valves which can be quickly cleaned. These pumps can deliver up to 800 lb per square inch pressure and are reciprocating, driven by air or steam on the principle of a single acting Worthington pump.

These pumps are so powerful that cases have arisen where continuous pumping at high pressures has caused damage to properties and lifted floors and parts of

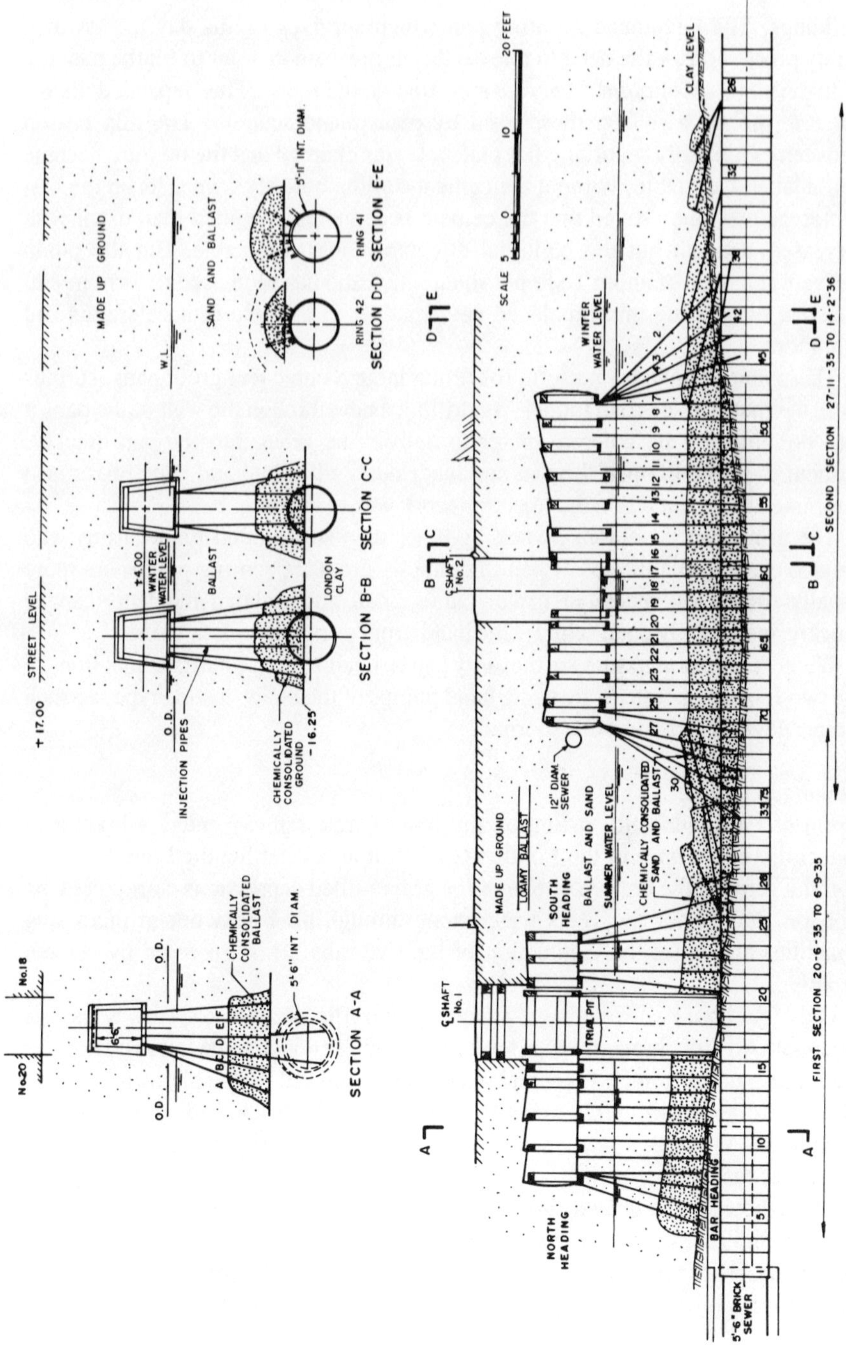

FIGURE 38. *Brook Green Sewer — Chemical injections from heading.*

buildings. The Greathead grouting pan which survives to this day has its own safety precaution — the need to release the air pressure in order to fill the pan.

In the 1930's *Colloidal Concrete* appeared on the scene. This depended upon a *colcrete* mixer, based on those used by paint manufacturers. The mix passed between two rapidly revolving flat plates. It was claimed that the mixture became colloidal and so stable, without the cement settling out. My own tests on the first prototype machine proved that the cement became better wetted than usual with increased strength but the colloidal effect was less noticeable. But the pump delivers at a head of about 10 lb per square inch and has been used to form in-situ concrete by pouring into preplaced aggregate, with the help of accelerators and inhibitors.

These pumps have been useful for filling large 8 cubic feet grout pans, surplus from the first Mersey road tunnel. By drilling a hole through the wall of the pans it was possible for the colcrete mixer to deliver the grout into the pan quickly, without the need to open the pan and laboriously fill it by hand. The mixer also pre-mixed the grout while the injection work was going on.

The grout pan is seldom owned by other than soft-ground tunnel firms who believe the sensible motto: "when in doubt — grout". Grouting pumps are more usually owned by specialist firms. Quite small and isolated incidents can be quickly and cheaply dealt with by the hand grout pump, not often seen.

Water stopping in fissures in masonry can be dealt with by alternate injections of the two Joosten chemicals by using hand pumps of the boiler-testing type, so such simple devices should not be scorned.

Farringdon Street, 1934

North of Farringdon Street Station the Inner Circle railway enters a bricklined tunnel driven *on bars* in the London Clay. But at several hundred feet from the portal it scraped the bottom of one of the gravel-filled depressions common to the London Clay. Over the years water came through the brickwork in increasing quantities and upset the conduction of the live rails. This we cured by cement grouting.

Only four hours were available at night after traffic stopped so speedy work was essential. An air compressor was located just small enough to be left inside one of the refuges in the tunnel. Plant and material was stacked on the platform. A plate layers's trolley was quickly assembled and loaded with scaffolding and drilling equipment. While this was erected the 8 cubic feet grout pan with cement and water was rushed in.

Many holes were drilled through the brick lining and fitted with expanding stoppers and gate valves. Holes revealed large gaps in the nine rings of brickwork where the bricklayers had scamped the filling of the bar-holes. Many tons of cement were injected over two weeks, but water still came through. Then work ceased over the Easter holiday. On our return we found the tunnel arch to be quite dry. The excess water had been given time to drain off and the cement to set.

Brancepeth Colliery

An unsuspected pre-glacial valley had been filled with soft glacial silt. A long access tunnel in sandstone had been used for many years, lined with steel arches and random boards. Unknowingly this had passed with only two feet of cover of rock at one place. One day a small block of rock dislodged and silt, described by the miners as *tooth-paste*, ran both ways up the adit. Luckily there were other means of exit.

I sent the mine officials to search the surface. The only sign was tilting of some fir trees. A man dug a four foot hole and then luckily, tested with a bar. He climbed rapidly out as a circular cavity, 24 feet in diameter was revealed, 120 feet deep.

The solution was to make the hostility of the material defeat itself. Two inch pipes were passed to the bottom of the cavity which was then filled with 3/4 inch shingle. A bulkhead was built in the adit for safety and the miners told to dig away the silt. This continued to flow until the shingle appeared in the adit. The ends of the pipes were then opened by small charges of gelignite and 100 tons of cement grout passed through them from a large grout pan, on the surface. This turned the shingle into a solid block and the adit was re-opened in safety (Champion 1951). **Figure 39**.

This is a warning of the hazards of glacial silt for anyone tunnelling for the first time in areas of glaciation. A few years later the press reported a mining disaster at Knockshinnoch in Ayrshire. There a large run of glacial silt made a wide funnel-crater to the surface and lives were lost. This needed more heroic methods to solve it.

Fly Ash and Woolwich Chalk Mines

The increase in fly ash accumulations at Power Stations after the war caused the Government to ask for suggestions for its use. Among other virtues it was found to have setting properties of its own and so joined the many materials which were becoming available for judicious choice.

A curiousity of South East England is the number of small chalk mines or workings, often encountered by chance. The chalk in much of the area is covered by heavy clay so, in the days when roads were non-existent it was cheaper to mine chalk for lime locally than to transport it for some miles from outcrops.

A housing estate had been built at Woolwich on Thanet Sand which extended down for 30–40 feet to the upper fissured chalk. An irregular system of galleries, 6 feet wide and 6 feet high had been worked as lime mines many years ago. Over the years the chalk had spalled upwards until suddenly it exposed the bottom of the

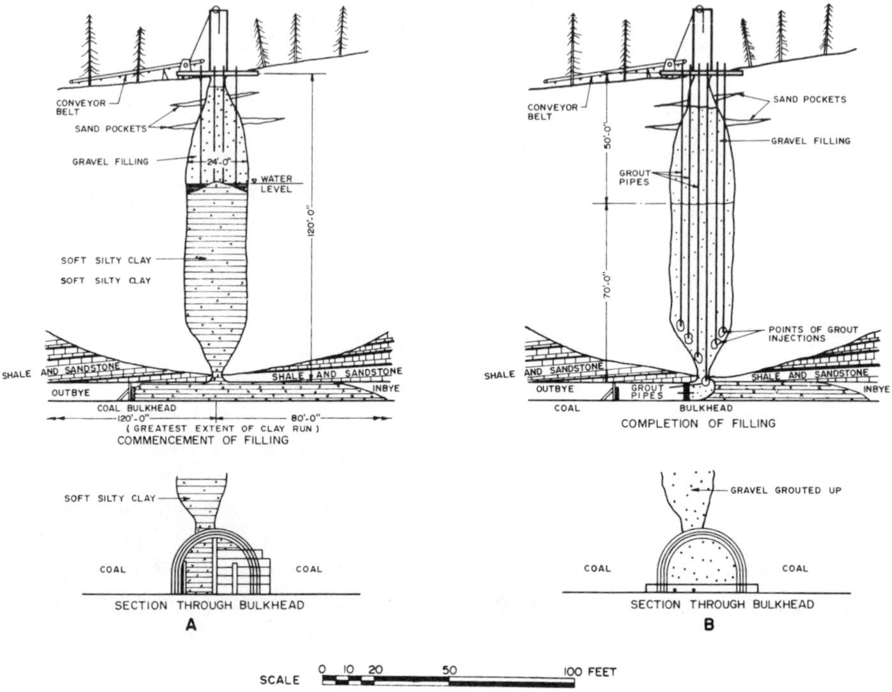

FIGURE 39. *Brancepeth Colliery — Treatment of silt run into haulage way.*

sand which naturally ran into the workings and caused craters to the surface, damaging the houses and engulfing individuals.

The solution which we adopted, described in *Grouts and Drilling Muds in Engineering Practice* (Butterworth 1963) was to fill up the galleries by flushing fly ash with water through flumes. The fly ash was placed in 4 feet lifts as the galleries had grown considerably in height. Sandbag bulkheads controlled the areas and the water drained off in the chalk fissures. The high points were grouted through pipes passed up through bulkheads with return-air pipes. Some of these grouting experiences could also have come under the next heading. Advice was also given on chalk workings found under buildings at Northholt and Bury-St-Edmunds.

Bo-Peep Tunnel

The early bricklined tunnels, both railways and sewers, are at increasing risk as time goes on. The most recent example is the collapse of the railway tunnel between Berwick and Edinburgh. The 100 year old Bo-Peep tunnel is between two stations at St Leonards-on-Sea. It was driven in Wadhurst Clay which comprises a variable series of clays, silts and sands. It had been lined with eight rings of brickwork but at the west end there was no brick invert, which was sometimes the custom at the time. The soil invert had heaved and cavities had developed in places behind the brick lining which was also in poor condition.

There is much to be learned by potential future tunnel repairers by reading the detailed paper (Campion 1951). The repairs were carried out by John Mowlem & Co. for which I was responsible as a Director, which meant visits to the site at intervals compared with the earlier works described which entailed daily or nightly visits. This work comes under grouting as this was an essential factor in the success of the work.

Traffic was suspended altogether and the invert concreted, reinforced with old rails, which come conveniently in railway work. Our major task was to build 170 partial rings of 26 feet diameter cast iron lining which sprang off cast iron bases on either side of the new invert. This involved cutting out the old brickwork but this was found to be in very bad repair with many lime mortar joints missing. Before cutting out the brickwork, we devised a massive grouting programme on an organised basis. The paper describes the gantries used to drill-in grouting pipes controlled by valves and many other details, as well as an interesting account of the soil mechanics considerations which followed the site investigation which had been carefully carried out (now that the Southern Region had their own laboratory!)

We used two grouting pans of 8 cubic feet capacity in which extra holes had been bored so that pre-mixed grout could be pumped into them by concrete mixers immediately they were empty. This saved the old fashioned method of feeding dry cement into a pan full of water through the hatch.

This gives the opportunity for repeating my own preference for grout pans rather than the high pressure grout pump in many situations. There is a safety margin as pressure must be released to refill the pan. I have met cases where damage has been caused by remorseless pumping of grout continuously.

This preliminary grouting was so successful that we were asked to carry it on beyond for two chains (as railway men are still calibrated) and this made it unnecessary to build any more cast iron rings. Most of the rest of the tunnel had been built with inverts or was in better or rock ground. A total of 645 tons of Portland Cement was used over a distance of 7 chains or 462 feet. The circumference of the partial arch was 60 feet and was 2 feet 10 inches thick.

GROUND-WATER LOWERING

Ground-Water Lowering is now the accepted title for forms of what are now called Geotechnical Processes, chiefly pumping from deep wells, shallow wells or well points. The latter is now the term to cover the jetting down of small diameter suction tubes connected to suction pumps as a rougher and more mobile form of shallow well work. Tunnels naturally provide less scope for these processes than for injections.

After I had served on the Code of Practice for Site Investigation, I found myself at the end of the second war appointed as chairman of a panel to deal with De-watering and Treatment of Sub-strata on the Code of Practice on Foundations. I wrote this section of the code myself with little alteration by my colleagues.

As I had put into it all my hard earned personal experience I took advantage of a request for a paper from some Local Associations of the Institution to write one entitled *The Principles and Practice of Ground-Water Lowering* (Harding 1946) which included actual examples. I believe that this was the first such paper to explain the whole principle to the "Unconverted natives" of the profession. It was printed in spite of paper rationing by the Northern Counties, Glasgow and Southern Associations but is now possibly hard to find.

There is now a generation of early middle aged engineering executives who have not learned the lesson of living through two World Wars and who accept the present availability of expedients with no thought to their evolution. So this may seem a good moment to set out the earlier customs and methods with which some of us had to struggle, as described at the Isle of Dogs.

Until and immediately after the 1914–1918 war, contractors, like the Army had to be mobile to carry out work in unexpected places, so that dirty but beautifully flexible power source, Steam, was almost universally used, except in cities if electric power was available.

Steel sheet piling did not exist so water was dealt with by sinking timber lined pits, reducing in size at each fresh setting of runners, while a small sump was being continuously sunk just ahead of the digging with some difficulty. Water was removed sometimes by that crude but useful steam Pulsometer pump which was almost self priming, or by Worthington-type double acting pumps. Centrifugal pumps came in with electricity which seems to have won the race over the petrol engine by a short neck. Diesel engines only seemed to be used in submarines.

In wide excavations the sump was usually sunk inside the work to save the expense of sinking separate sumps outside so water was drawn into the work. This way of going on often led to loss of ground leading to subsidence. A common accusation could be heard of *drawing sand out of the ground*. The whole principle of Ground-Water Lowering is to draw the water away from the work.

This was the state of affairs at the Isle of Dogs. In most firms, today's mechanical department was called the Black Gang, run by the ex-Steam boss, whose job seemed to be hereditary. Ours had little sympathy or experience of petrol driven pumping plant. Then when the agreement was made with Siemens Bau-Union through Siemens Schuckert Great Britain Ltd., I was able to take this out of his hands and deal with it myself.

Our first use of the methods of deep and shallow wells were not on tunnel work.

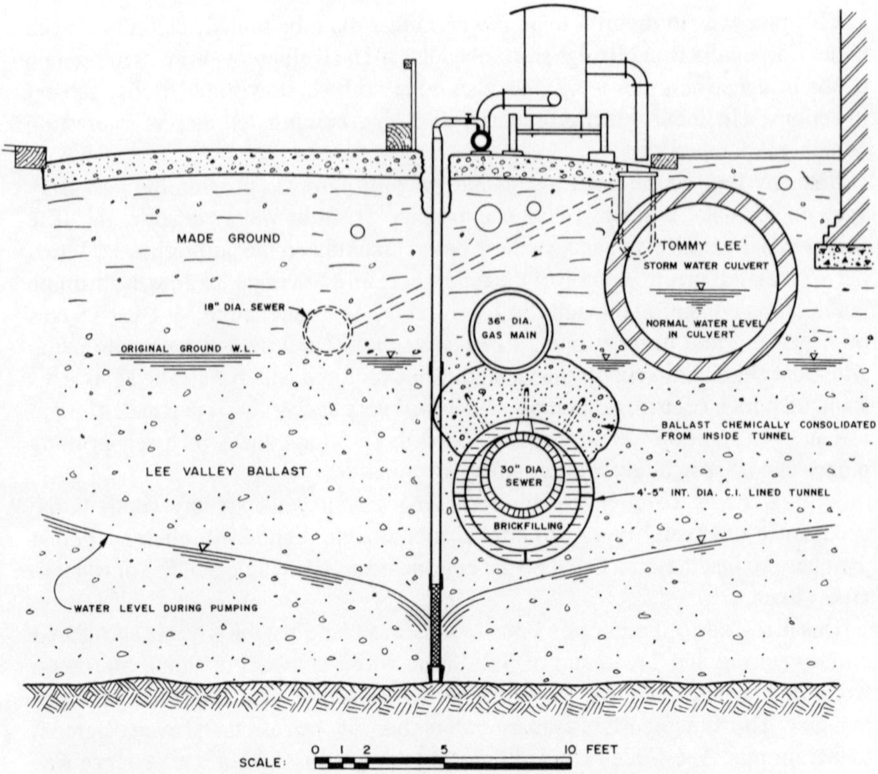

FIGURE 40. *Sewer diversion under gas main using well-points and chemical injections.*

It was not until I became Agent on the Bow-Leyton extension of the Central Line that opportunity arose. About this time word came of the development of well-points in the U.S.A. but on a less scientific basis than Siemens methods. Apart from theory propagated by Dr Sichardt and others the strong point of the Siemens method was that no sand was lost when pumping from bored wells equipped with gravel and mesh filters.

Well points were jetted into the ground with plenty of water and the necessary high pressure in a rough and ready manner. I was very anxious to buy the first ever Moore-Trench equipment when it was offered in this country but there were fears that this would clash with the agreement with Siemens. However, I was able to buy the second set and use it on a long trench to carry a 30 inch sewer diversion on the Bow-Leyton work. **Figure 40** shows how it was used on a section which had to be carried out in tunnel by means of a 4 feet 5 inch cast iron lining driven diagonally under a large gas main and the large Tommy Lee storm water culvert. Injections were made ahead by the Joosten process from inside the tiny tunnel.

This pipe was to divert a main sewer so that the tube tunnels could be driven under Carpenters Road Bridge, described later. The well points were used right up to the bridge where the work was also done with shallow bored wells, cement injections, chemical injections, bored piles and compressed air, an interesting collection of expedients.

Siemens Bau-Union used a separate vacuum pump to provide the necessary vacuum to make a series of inter-connected shallow wells self-priming. The Mooretrench pump also had a vacuum pump mounted on the pump chassis. I tried out a Sykes self priming pump of the type where an extra impeller does the priming but this was not strong enough. So I showed the Mooretrench to Mr Paish, Sykes Manager and told him I wanted a British version. He provided a vacuum pump with float chamber to cut off the vacuum when required which was fitted above his standard 6 inch centrifugal pumps, which was very successful. I persuaded him to use an even more powerful vacuum. Now these 6 inch and 8 inch self-priming pumps can be seen all over the country.

A great boon to engineers for removing casual water at any depth is the appearance of several types of portable submersible pumps with their suction at their base so that they can be set on the bottom of sumps or large puddles of water to great effect.

This first Code of Practice on Foundations was compiled thirty years ago. Part 4 on Dewatering and Treatment of Sub-Strata was a compact textbook of 10,000 words on the State of the Art at the time but on re-reading there does not seem much to regret. The thought arises as to the extent that anyone read it and even offered it to their pupils. The Code included Freezing though this Author's experience was limited to two unsuccessful tenders for freezing shafts at Swansea Power Station and Mouldsworth.

Another expedient was not included in the Siemens agreement but was an offshoot of it. We found the Pressure Piling type of tripod and especially the Ingersoll air winches useful for sinking deep as well as shallow wells. After the War, when steel as well as paper, petrol, food and clothing were closely rationed it was difficult to collect enough steel for lining tubes. We insisted on using square threads to prevent damage and kept to short five feet lengths. The deep wells were bored with 20 inch diameter to give room for twelve inch lining tubes to allow for four inches of gravel filter with room for the submersible deep-well pumps. To overcome friction we ordered 24 inch tubes for use in the upper section if needed.

The fact that we had 24 inch tubes made it possible to come to the aid of a friendly rival by boring to that diameter when bored pile firms could not go beyond half that diameter. So we found ourselves entering the bored pile field with the first advance in size which later reached such large diameters. One of the revolutions of the Macmillan Age was the change of heart among firms to actually spend money in development in place of their pre-War Dickensian Scrooge-like thrift.

National Physical Laboratory Bomb Disposal

During the war contacts with the Royal Engineers increased. In one case an unexploded bomb ended its swerving path below ground under the only tank for testing ship models, at the National Physical Laboratory, this caused concern at 10 Downing Street. The bomb passed through twenty feet of Thames gravels containing ten feet depth of ground-water, before coming to rest in the London Clay thirty four feet below ground level. A Bomb Disposal Section sent to deal with it were given a set of Moore Trench Well-point equipment. As they had no instructions how to use it they put all the well points into one four feet square hole to no effect. As, three years before, I was one of the first people to use this plant I was sent down to advise. The first thing to do was to jet down the well points with plenty of spread around a hole twenty four feet by eight feet. The clay was reached with two settings of timber runners but, naturally, there was about a foot of water over the clay which will always pass the points when they reach impervious ground. This discouraged the R.E. Officer but I told him to put on his gumboots and dig out a grip for a sub-drain under the water along the foot of the runners to a sump to be dug in one corner. When a snorer flexible suction was connected to the well-point suction main and dropped into the sump the water disappeared at once. I then insisted on the sinking of a box-timbered shaft to a depth which would give at least five feet of clay above the top of the heading which then could be driven along the trace until the bomb was reached. One precaution was to keep the timbering two feet above the clay so that men could get out if the pump broke down. **Figures 41a and 41b**. This all worked out so well that the R.E. wanted to paint my mud guards red to match theirs but I did not think that my firm would approve.

From this I was asked to give a one-day teach-in to Bomb Disposal Officers on the use of well points and problems of excavation. This took place in the Duke of York's Headquarters, which by then was entirely windowless. My firm decreed that I could be contacted at any moment to give advice free of charge if needed. This led to a number of interesting contacts with Major Yates and his merry men. Although the Army must have enlisted many timbermen and tunnel miners it was surprising how few such talents were provided for the gallant Bomb Disposal Sections. At times they were in more danger from their own excavations than from the bomb they were seeking. I then served with Sir Andrew MacTaggart and W.H.G. Roach as a Committee of the Federation of Civil Engineering Contractors which drew up a code with drawings of a standard form of timbering for the Bomb Disposal Units.

An early form of well-point is the Abyssinian tube well used by the Sapper General Lord Napier on his march to Magdala in 1868. Each consisted of perforated pipe with a pointed shoe. The pipes were driven into the ground by sledge hammer. If water was found they could be fitted with the old fashioned village hand-pump.

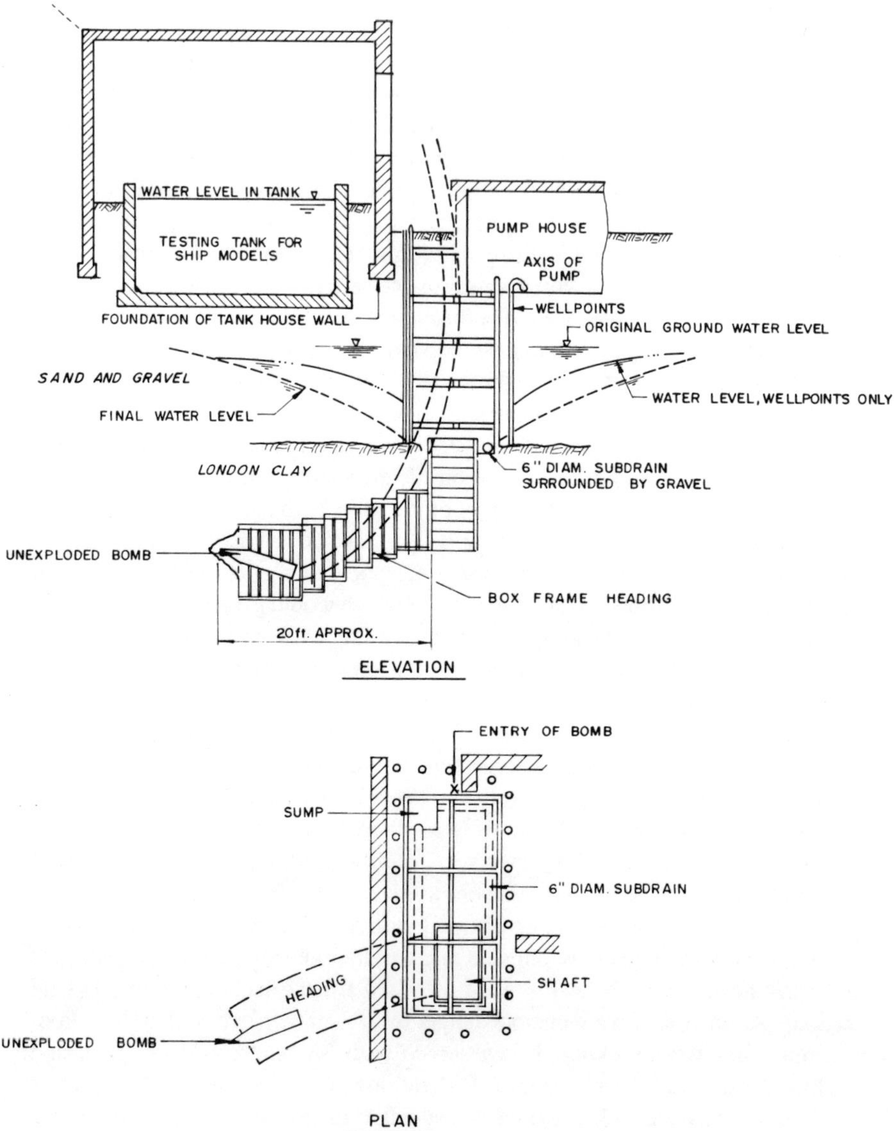

FIGURE 41a. *Well-points for bomb disposal under National Physical Laboratory.*

Two cases occurred when mending bombed sewers where to our surprise well points failed to perform their function, until sub-drains were unearthed below the sewer and were stopped up. On first asking the young L.C.C. Main Drainage Engineer if they existed under the sewer he replied "What is a sub-drain?". So the recounting of old fashioned methods can be justified.

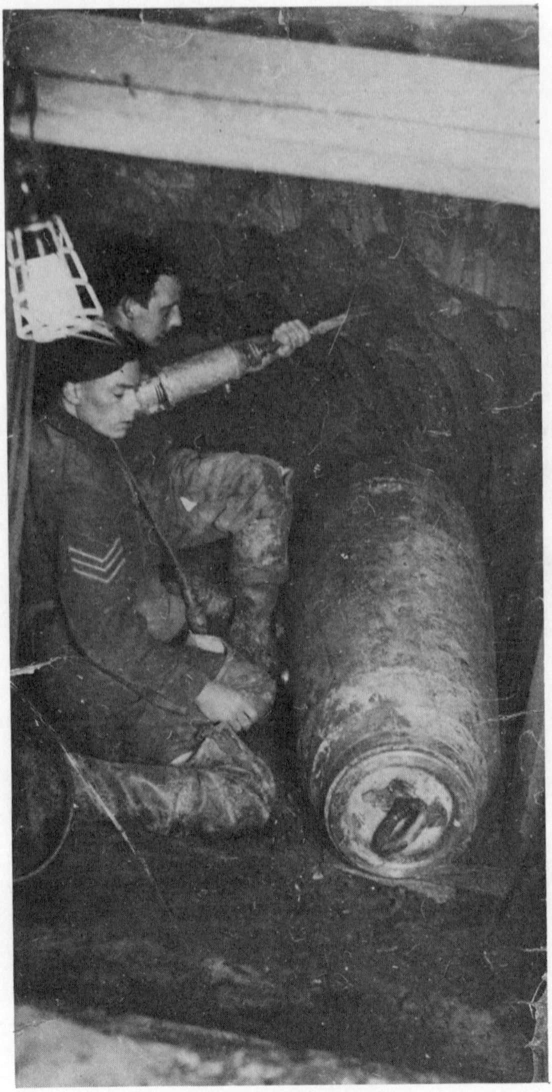

FIGURE 41b. *Unexploded bomb fifty feet under National Physical Laboratory.*

After the war, additional escalator tunnels and passages were required at Charing Cross at the time of the Festival of Britain in 1951. In contrast to the primitive methods which we used in 1925 eight deep wells were sunk on either side of the station in a yard under the Charing Cross Railway bridge which lowered the water in the Thames gravels. Other examples are mentioned in the age of advising.

Chapter 8

SETTLEMENTS AND CAVITIES

In *From Theory to Practice in Soil Mechanics* (1959), the collection of the works of Dr Terzaghi, there is a section on his work in 1938 on the Chicago Subway by Dr Peck. This describes the steps taken to predict, prevent and measure settlements in that notoriously soft clay in the area.

This is a useful comparison with the entirely different behaviour of the stiff fissured London Clay at depth; though where it outcrops the clay has been weathered and is soft. It was the custom to take photographs of buildings along the line of future tube tunnels before the work started in order to counter false claims, some of which were made before the tunnelling started! In the early London tubes few scientific measurements were taken.

I do not recall any trouble over settlement on works on which I was engaged. Settlements must have been small for several reasons. By law the tube railways, at depth, had to follow the street lines, except by special dispensation. The clay was stiff. On well conducted contracts the lining was grouted at every ring and the face boards often grouted at weekends.

The streets were riddled with trenches for services and sewers and the road surfaces were not up to present standards. Houses with shallow footings on clay were more liable to disturbance by weather than the tunnels. With large buildings under construction it was usual to transfer benchmarks on to the structure for levelling as soon as possible, so comparison with outside levels passed unnoticed. It is only in recent years that sophisticated means of measuring distortions of linings and settlement records have been adopted.

Cavities formed by collapse, runs, or *blows* in a tunnel during driving are a different matter. Those of us who were well brought up under good and experienced masters had to await the experience of observing such collapses when called upon to rescue the works of a rival, or when encountering cavities left behind, unmarked by their perpetrators.

Much profit can be gained from reading. I had just read in the *Life of Sir Francis Fox* how he had encountered a spherical cavity in the Thames sand and gravels. He deduced that this was caused by careless driving of a sewer at a lower depth many years before. A few days later we encountered just such a spherical cavity under a tram-line crossing when driving a timbered heading to carry high voltage cables in Shoreditch between Lyons and the Unicorn Public House. This was over a deep level sewer below, so was a legacy of the previous miners.

Since then other cases have come to my notice, often when the tunnellers have emerged ususpectingly into one of the many gravel filled depressions in the eroded surface of the clay which seems to appear constantly in these notes. The ability of the sands to arch-over without reaching the surface has led to some cavities being undetected.

SUBSIDENCE UNDER UNIVERSITY OF KENT

In 1975 I was asked by Messrs Harris & Sutherland to add my opinion to that of others on the collapse of a disused tunnel after 150 years' life. This was the first British true railway tunnel driven during 1827 through London Clay in the affected area, then passing into the lower Oldham Sands of the Woolwich and Reading Beds.

This small tunnel was 771 metres long, with almost vertical side walls standing on a brick invert with a 4 metre diameter brick arch of four courses. Traffic had ceased in 1952 but the tunnel was still maintained until 1964 when the land above it, including the tunnel, was bought for the building of the University of Kent. Bulkheads shut off through-ventilation but doors allowed access for inspection.

In 1973 the Architects reported that "the section which had deteriorated most seriously lies under the Cornwallis Building and this makes it imperative that some action be taken as the tunnel must not be allowed to collapse". Unfortunately time was spent in making a decision and while tenders were being prepared in July 1974 a local collapse did occur, and part of the Cornwallis Building was damaged by settlement and had to be demolished.

The collapse occurred before the various parties consulted visited the site. But from all the available evidence the general consensus led to opinions that a failure of the brick invert caused the walls to fall by bending so that the arch collapsed and blocks of clay filled that length of the tunnel. A soils investigation under Professor Bishop with the consultants, Harris & Sutherland, indicated that the ground failed in undrained shear, forming a single plug which moved downwards into a cavity created by local failure of the tunnel lining and surrounding clay.

There is a letter from Sir Marc Brunel dated October 17th 1836 to "My dear Isambard" in which he writes at length on the shape of linings. He alighted near Daventree to examine a slip in a new tunnel and he drew a section identical with the Kent tunnel and showed where the side wall had collapsed and the tunnel filled entirely, at the place where the ground broke in. He writes "They have not, I conceive, given sufficient arching to the sides as here shewn. This outline would have been preferable".

This was somewhat egg-shaped, as Isambard used on the Box Tunnel. It is remarkable that the Kent tunnel stood for so long.

In my report I drew comparisons with other collapses from my own experience and quote parts of it: –

Comparisons With Other Types Of Failure

When tunnels are being driven in London Clay, if the tunnel face and crown are insufficiently supported, large blocks of clay can fall out of the face or roof leaving cavities overhead. But in such cases the cavity can be quickly filled as the miners are present with all necessary resources. So the behaviour of clay in reacting to such cavities is not often on the scale at Cornwallis and not often observed.

"As an example in my own experience, a bomb dropped in Trafalgar Square, shattered the heavy cast iron lining of the 26 feet diameter chamber at the foot of the escalator but did not penetrate. The lining dropped on shelterers, followed by massive lumps of London Clay leaving an arch-shaped cavity above. But the miners were available next morning who were able to erect supports and fill the cavity before any further effect followed.

"When a bomb shattered the heavy lining of the Balham Tube Station down to platform level the London Clay above dropped as a solid block. As the top of this was below the crown of the tunnel, the water from burst water mains and the ground-water, washed in sands and gravel which buried the shelterers. As miners worked to clear the tunnel the clay was so intact that it looked as if the miners were excavating a normal clay face.

"The case of the railway tunnel at Bo-Peep near Hastings is described by Campion, J.E., (1951). The invert started to rise and the side walls settled 7 inches. The movements were over several years and cracks appeared in houses over the tunnel. The final observations showed a *draw* at the surface for 30 and 35 feet on either side from the centre line of the tunnel, which had an internal diameter of 22 feet with walls 2 feet 10 inches thick. Although there was considerable cracking the arch was supported by heavy timbering prior to lining with cast iron.

"In the discussion on the paper I described an experience at Brancepeth Colliery where glacial silty clay had flowed into a mine gallery. The result was a cylindrical cavity reaching 120 feet to the surface. A student at Cambridge University described in a competition his research by models of the similar effect of unlined tunnels in soft clay. Even in small scale he observed a settlement of almost vertical sides.

"The seven feet diameter Tower Subway was also crushed by a bomb which fell in the River Thames but without penetrating the tunnel. In that case the clay seemed to have gradually squeezed the fractured lining without forming a cavity (Harding 1945). **Figure 42**.

"Many ways in which the clay could change its properties were discussed, with possible cavities left by careless driving and lining, change in ventilation and the effect of water. The brickwork could have been improved by careful grouting but

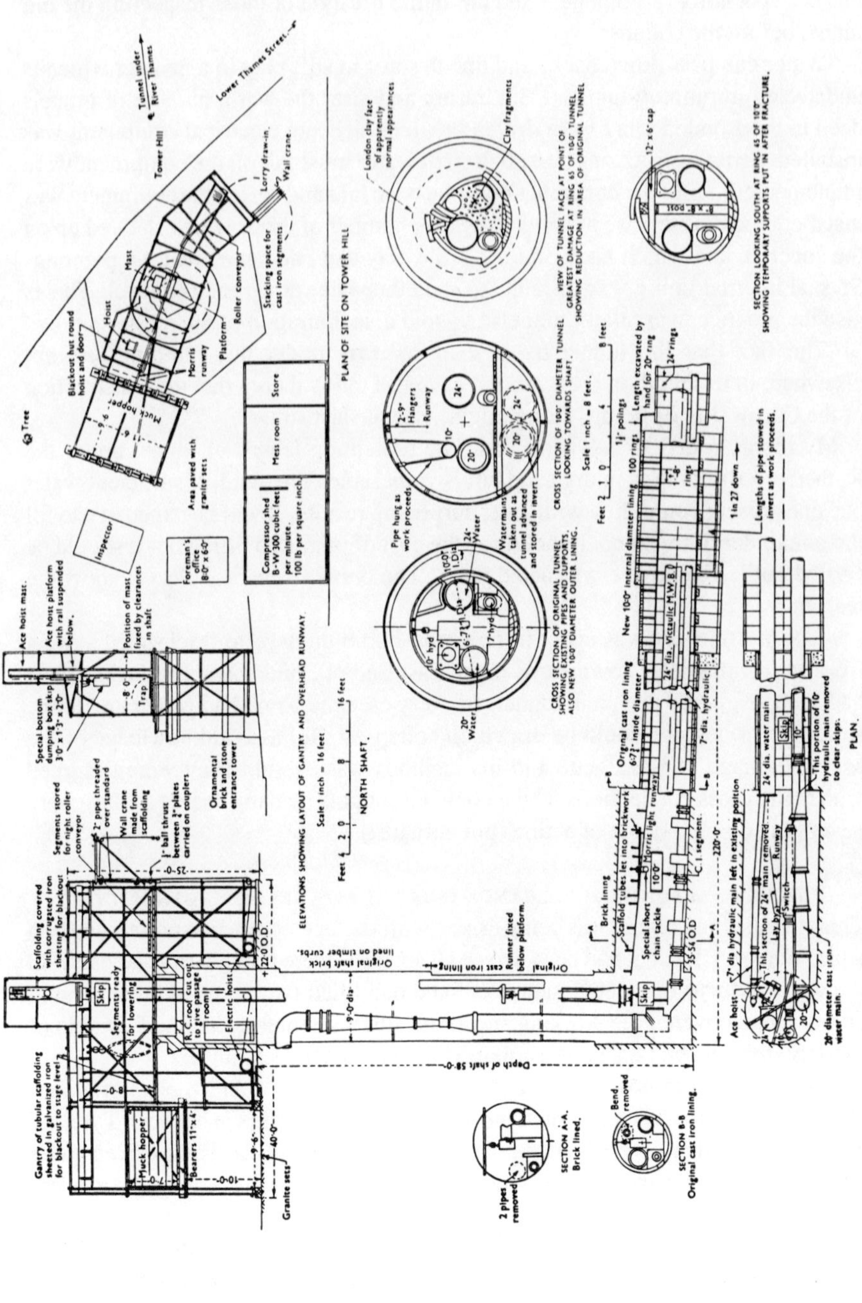

FIGURE 42. *Repairs to Tower Subway after bomb damage.*

this is a speciality of tunnellers and not in the lifestyle of those inspecting the old tunnel before the collapse.

"Water can pass down backs and fine fissures in stiff clay in a manner which is undetected in running tunnels. But during and after the war a number of tunnels deep in the London Clay were driven in which delicate electrical equipment was installed. Strict instructions that no drop of water must fall on the equipment were implemented. After the concrete floors had been laid and before the equipment was installed it was a surprise to many to see the number of drops which showed up on the concrete and which had come from the clay and had penetrated the pointing. Special internal lining was installed to shed the water down to floor level. This is also the practice in escalator tunnels to avoid discolouration of the plastering.

"The fact that the tunnel *decided to collapse* under this building and not elsewhere in its length in open ground gave rise to my theory that the construction of the Cornwallis Building, though light, was the last straw".

My report ended by advising that "the remaining length of tunnel under the Rutherford Building is in dry sand but is vulnerable. The sand, if sufficient water became present could flow with quite surprising results. It will be expensive to fill the remainder of this tunnel, including the cost of imported filling but it should be carried out". The advice was acted upon. The previous recorded cases show the reason.

Most of 1940–41 was spent in repairing bomb damage to brick-lined sewers from 12 feet diameter downwards in the East End of London and the docks. Most had been originally driven in tunnel. In every case the bomb had penetrated to the sewers so no lessons could be drawn on collapses. But a good knowledge of the Main Drainage of Bazalgette and his methods was useful. They were designed with many cross connections which could be shut off by dam-boards so they were never put completely out of action (but only just).

THE AWALI TUNNEL IN LEBANON (FOR THE LITANI RIVER AUTHORITY)

This ten mile long tunnel was being driven with six faces, from two portals and two adits. When 67 per cent had been driven all six faces were brought to a standstill. In the worst case 70,000 cubic metres of sand had filled two miles of tunnel and in addition had covered the working site at the adit with three quarters of a metre of sand. So this can be treated as a collapse and not under rock tunnels.

The Litani River flows south between the Lebanon and Anti-Lebanon mountains along the rift valley and then turns west to flow into the sea between Tyre and Sidon. In 1959 dams were being built and the Awali tunnel was being driven under Mount Lebanon westward to provide hydro power and then to irrigate the western lowlands.

The work was sponsored by the World Bank, controlled by the Litani River Authority in Beirut, engineered by a French consortium of consultants and contracted for by a Franco-Italian combine. The tunnel, supported on steel arches was horseshoe in shape, 3.24 metres in arch diameter and 3.25 metres high. The folded and faulted strata are mostly Cretaceous or Cenomanian with a syncline between the east portal and the next, Jezzine, adit which produces a limb of parallel beds rising from west to east up the mountain at 35 degrees to the line of tunnel.

The drive from the river portal by the Markabi dam was driven through Jurassic fractured limestone, and after encountering a flow of water of 6,300 gallons per minute, eventually blocked off, came to a standstill in bad conditions after 3,587 metres. Because of the other events the Authority ordered advance to stop and to begin concrete lining.

The drive eastward from Jezzine adit had passed the syncline of Cretaceous dolomitic limestone and then into brown marls at this 35 degree slope without trouble. Then after a narrow bed of limestone it reached the junction bed between this and the poorly cemented Cretaceous sandstone. This is when the break occurred.

Personal experience then comes into play when invited to join a World Bank Commission in March 1960 to advise the Litani River Authority. This was led by J. Donovan Jacobs, of Jacobs Associates of San Francisco, with J. Kell — Senior Partner of Mott Hay & Anderson, Consulting Engineers, Mansell MacLean, the Contractor for the Lincoln Tunnel and Victor L. Stevens of Boyle Bros Drilling Co, Utah.

Our terms of reference were to study the physical difficulties encountered in tunnelling through the cretaceous limestone and to recommend steps which should be taken to insure the completion of the tunnel. The thirty page report was typed on the spot with the help of a secretary borrowed from the U.S. Embassy, after eleven days of intense work with inspections and meetings to gain details. Extracts from the report may shorten the description:

"The difficulties which have arisen, and which have disrupted the progress of the tunnel, are entirely associated with the Lower Cretaceous Sandstone which must be traversed in various sections of the route. A total of 67 per cent of the tunnel length has already been driven through the limestones, marls and volcanic material which constitutes the greater part of the strata. This part of the tunnelling work appears to have been carried out expeditiously with the use of appropriate methods but, unfortunately, without adequate and properly conducted forward exploration, particularly when approaching the vicinities in which important changes in the strata were to be expected. ..."

From the section *History and Present Situation* under *Jezzine Upstream*:

"... This was the situation on 3 August 1959. When the face had reached station 3,207 a small flow of water, probably about 5 litres per second, was running in through a lenticular opening in the lower left-hand quadrant of the face. Contractor had drilled 7 holes in the face to a depth of approximately 80 centimetres. These holes were loaded and blasted. After the blast, and a short period of ventilation, the workmen returned to the face to discover water flowing in through the pile of broken rock in the bottom of the tunnel. An effort to construct a barricade with steel beams and plates in the vicinity of station 3,201 metres proved unsuccessful and was abandoned when the inflow of sand reached that point."

The sand then advanced in a series of spasms until, after an attempt to dig back towards the face, the final rush filled the tunnel and washed out over the area around the adit.

There must have been a large amount of water accumulated in the parting plane between the limestone and the sandstone. The mountain is 1,000 metres above the tunnel level where the sandstone outcrops and some opinions considered a possible head of 500 metres. The sandstone in a quarry could be seen standing vertically for a considerable height and was quite hard on the outcrop on the mountain top. **Figure 43**.

Among the documents examined was a report from Monsieur Goguel, Head of the French Geological Survey and joint author of our paper on the Work of the Channel Tunnel Study Group 1958–60. This led to these paragraphs in our report:

"Among the geological reports, that of Monsieur Goguel of 1956 calls attention to the gradient of hydrostatic pressure in the sandstone and the possible existence of soft clay bands. He also gives warning that the gradient can reach a point where it exceeds the mechanical resistance of the grit. This is a clear warning of the possibility of internal erosion if care is not taken.

He concludes that *in brief* the erosion of the Lower Cretaceous and the passage to the Jurassic seem to produce a short and very difficult passage.

This is a clear warning given by a geologist of great reputation, and again points to the fact that Jezzine upstream was at a point where great care was called for at the time of the sand run, which is a case of internal erosion.

Several times in the French reports, the importance of exploratory drilling ahead is emphasised.

If this advice had been acted upon, the accident might not have occurred."

Our report contains a section on Exploration and Drilling Methods at some length. Some points were:

"All holes should be cased so that drilling can be conducted through a packer gland. The packer should have a pressure gauge for testing water pressures. All exploratory holes should be started sufficiently large enough so that it is possible to case off in bad ground, reduce the size and continue the hole. It is therefore, important to have several sizes of casing on the job so as to ensure the maximum depth of exploration.

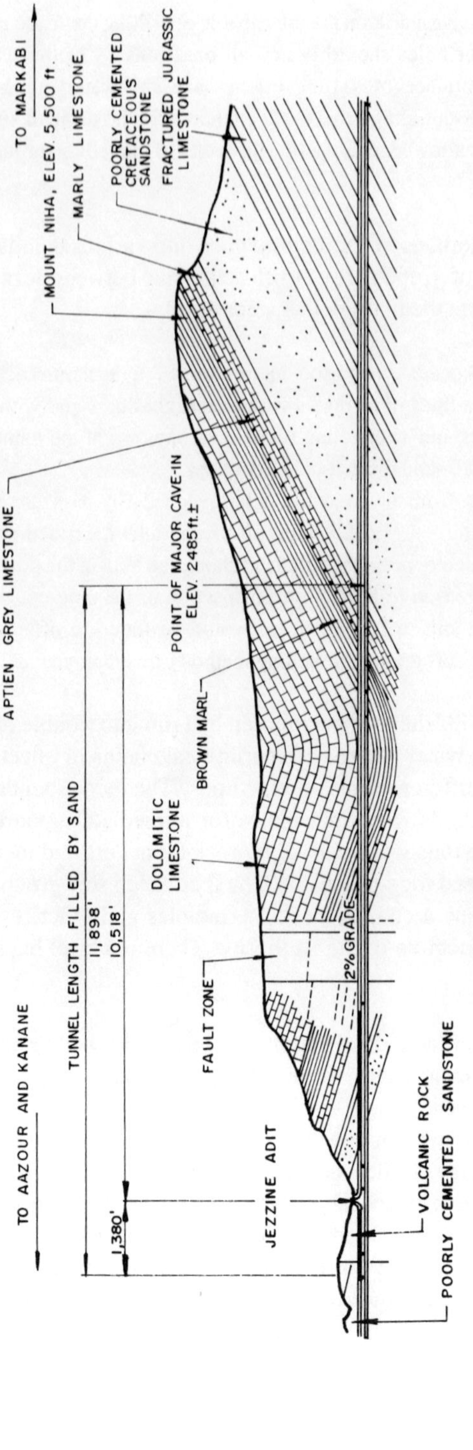

FIGURE 43. *Alawi Tunnel, Lebanon — Geological profile.*

It would be wise to have a drill on the job capable of drilling up to 300 metres ahead of the tunnel face. The feeler holes should be, on all occasions 15 metres ahead of the tunnel heading ... The importance of sealing off at once any water met with in borings is re-emphasised. It is essential that the water contained in the difficult strata be kept in the static condition. If it is allowed to move, the difficulty of tunnelling through the ground may be greatly increased."

The section on Methods of Tunnelling goes into various details in some length. For the major drive of 3,566 metres to close the gap between Jezzine and Markabi through the sandstone the main points were briefly:

"To build safety bulkheads every 300 metres or so, to reclaim each advance. These bulkheads to be on the lines of air locks with watertight doors and with 1 metre diameter overflow by-pass pipes and a collecting funnel near the roof of the tunnel. A smaller pipe would carry the pumped sand discharge. **Figure 44**.

When a safe distance from the break through is reached a by-pass tunnel should be driven to circumvent the disturbed area but with all precautions described and by soft ground methods. A strong concrete pressure bulkhead should be built at the point of turning.

There seems good reason to think, from inspection of the outcrop, that a considerable portion of the ground still to be penetrated would be found to offer no great difficulty provided that ordinary soft ground tunnelling methods are employed ..."

The other faces with shallow overburden had run into trouble due to collapse of the sandstone when water entered. Experimental chemical injections were being carried out by two different firms and methods. The recommendations here were for soft ground methods with liner plates for immediate support and with a top heading. As the previous work was all by rock tunnellers and methods, emphasis was placed on the need for supervision by experienced soft-ground staff.

The report contains a chapter on the principles and practice of grouting and injections for the education of the Authority. There is a brief but strong comment on organization:

"One obvious aspect of the present unfortunate state of the work is an apparent lack of co-ordination and communication between Owner, Engineer and Contractor. This is a matter of grave consequence ...

We cannot too strongly emphasise the absolute necessity for unified direction of a kind that will enable the many practical decisions which must be taken to be put into effect at once without confirmation by several persons or bodies sharing divided responsibilities ...

Experience in the performance of difficult tunnel operations has taught us that in this class of work, perhaps more than any other, it is vital that operational control should be in the hands of an experienced and trusted man whose on-the-spot decisions are implemented

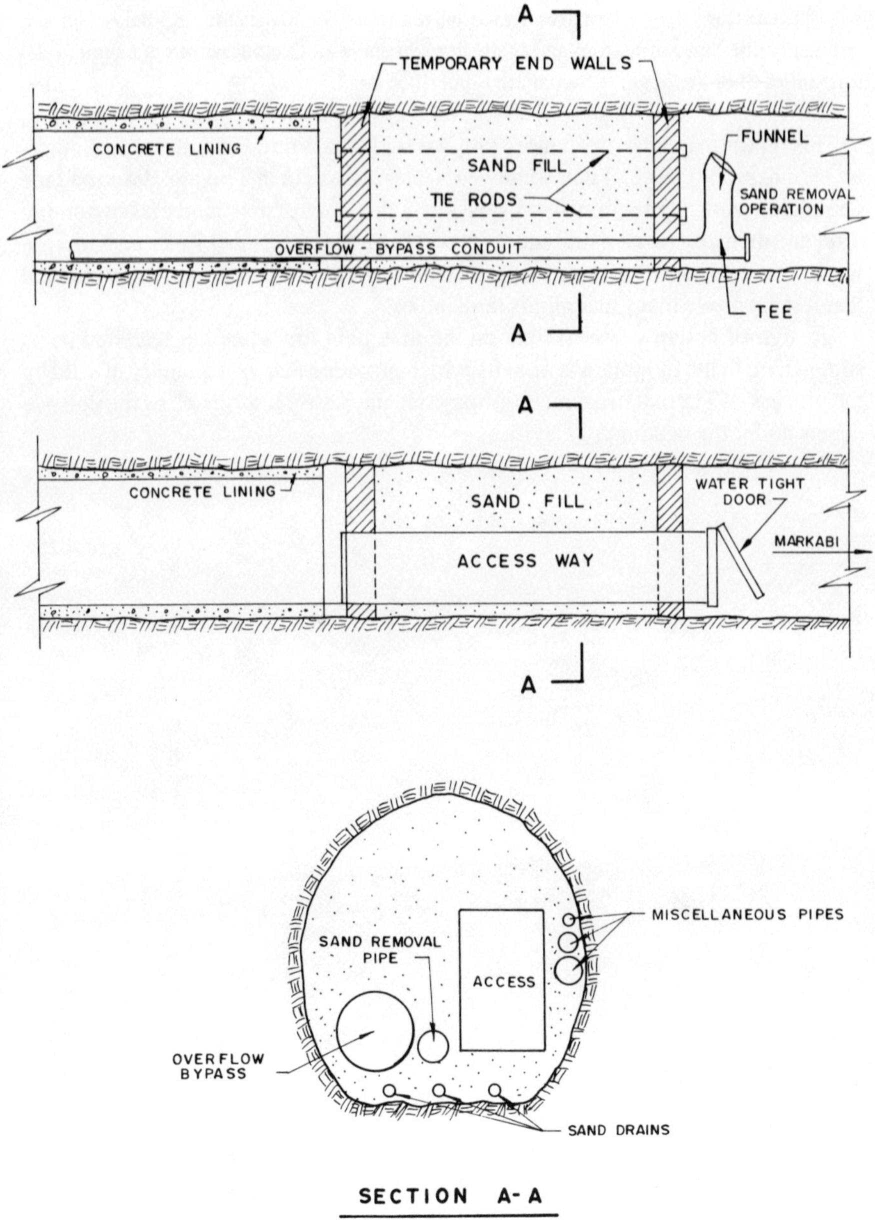

FIGURE 44. *Alawi Tunnel, — Proposed safety bulkheads.*

without question. Any alternative which leaves room for indecision and delay can and frequently does have disastrous and costly consequences ... Considerations of payment and formalities of protocol are secondary in times of crisis."

This report may have been more popular with the World Bank than the Authority. We heard no more. Then some years later an article in Travaux described the completed work. It mentioned our Mission with names of the members but nothing else, and then described a subsequent report from Eléctricité de France and how the work was carried out on their recommendation. The similarity is so co-incidental that it must prove that great minds think alike!

No sign of collapse was visible on the mountain top when we searched it, so somewhere in the mountain is a cavity which provided that vast quantity of sand by the process of internal erosion. This may one day cause a surprise, so the episode comes under the heading of Cavities!

Chapter 9

MORE WORK IN COMPRESSED AIR

CENTRAL LINE EXTENSION, BOW-LEYTON (1936–1940)

This work is described at more length by G.L. Groves (1946) in his paper on the Ilford Tube. **Figure 45** shows a simplified section of the ground. Ten hooded shields with water-traps were used to drive twin tunnels of 12 feet interior diameter for 10,000 rings. The west section, from two working sites was driven mainly under an embankment carrying eight tracks of the Great Eastern Railway, so considerable precautions were considered necessary.

For the first time the Joosten two-chemical process was actually specified. The beds of the Lea River tributaries were solidified from floating pontoons. In addition, at the Waterworks River, where cover was only 7 feet, a 7 feet hooded shield, built inside the 12 feet main shield, was launched and drove a pilot tunnel for each line. Extra grout holes, large enough to allow injection pipes to pass were used in the lining. This was the forerunner of later developments at Dartford and Blackwall. From the pilot tunnels the ground under the abutments and central pier was also solidified.

The water traps on the twelve feet shields were very unpopular as they restricted entry and made shovelling out difficult. The shields for the seven feet pilot tunnels had to be driven with no protection under the Waterworks River with only a few feet of cover. I fitted them with traps by bolting semi-circular steel plating, stiffened by angles to the ends of the shield jacks which fortunately had bolt holes. The lower half of the timber grouting ribs were bolted outside the steel plates so after the shield advanced the grouting rib was complete, and there was plenty of room in the improvised trap for mucking out. If such traps are again in fashion this method of a semi-circular plating outside the shield jacks might be an advantage on the full size shields.

On this contract, the Foreman, Joe Trelevean devised a simple form of shield-plough to prevent rolling. This meant cutting a slot on either side of the shield skin at axis. This coincided with the cross flanges of the shield stiffening rings so a steel plate could slide on the flange, secured to one of the bolts and could be pushed out or retracted as required. This recalls the saying of Sir Benjamin Baker when asked "Why should a shield roll?", he replied "Why should it not?".

When Dalrymple Hay first tried to work his idea of a hooded shield with clay

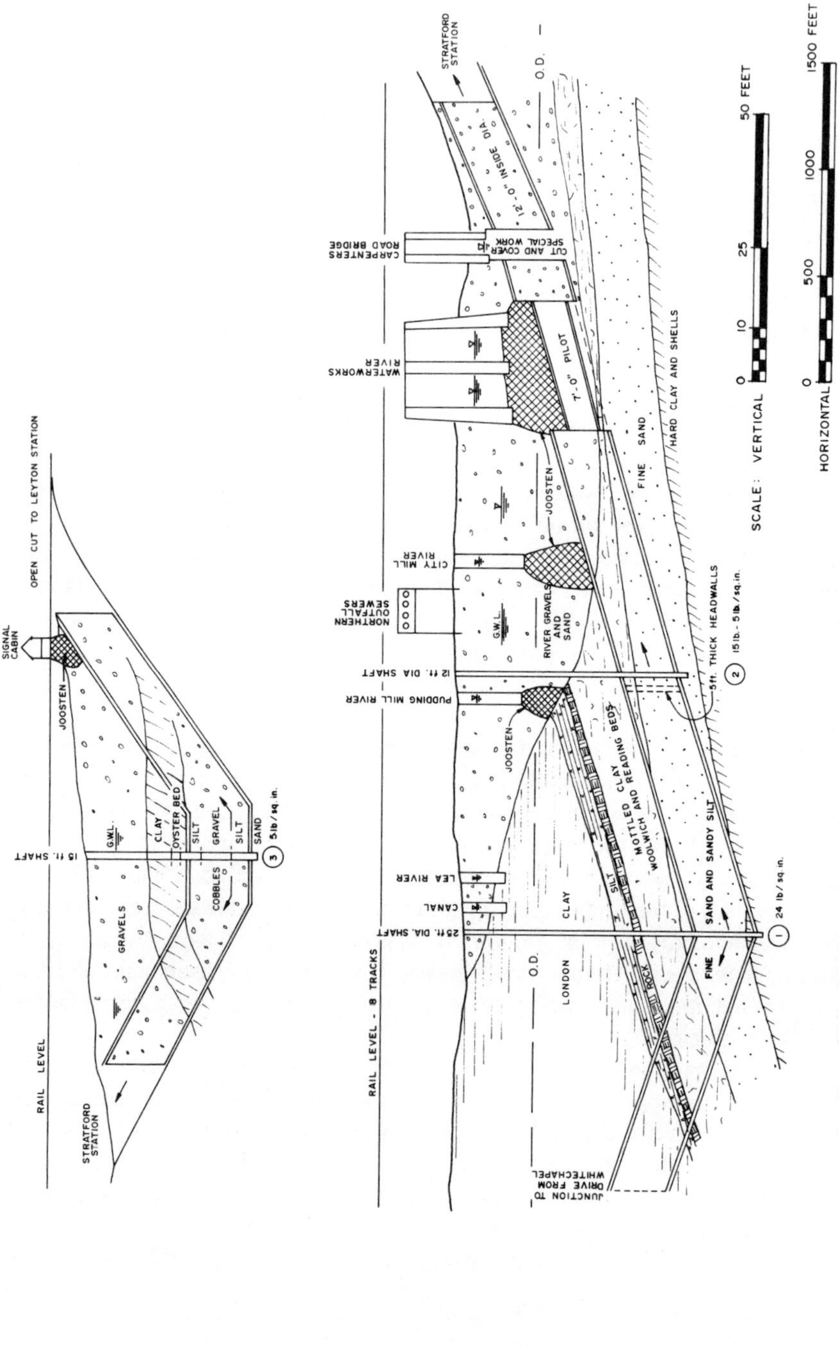

FIGURE 45. *Central Line extension, Bow Road to Leyton — Longitudinal section.*

pocketing on the Waterloo & City railway the miners all walked out. But Joe Trelevean told me that as a young miner he stayed behind and he worked round the left half of the face while Dalrymple Hay worked on the right and so confounded the sceptics.

All the injection work was completely successful and the shield driving proceeded with almost monotonous regularity. Other events provided variety. There were three complaints from old paint works, close to the small rivers, that the escaping air had entered their decrepit cellars. After the wartime bombing all three buildings had ceased to exist but the lawyers were still arguing with the Transport Board over compensation!

The fine sand with silt partings at shaft 1 required 24 lb per square inch pressure and became so firm that spader blades wore out after 350 cubic yards of digging. But if the pressure dropped one or two pounds the face became in danger of complete collapse. Only 5 cubic feet per minute was needed per square foot of face.

Clay pocketing was used in the sands and gravel and 20–30 cubic feet per minute per square foot of face was needed even with caulking carried close to the face except where the ground had been treated.

The drive from the east shaft beyond Stratford Station was in typical Woolwich and Reading Beds of sands, cobbles, clay beds and thin beds of limestone. The original site investigation had reported no water. In fact it turned out to be fully water-bearing requiring 5 lb pressure but otherwise was uneventful except for one lesson: do not rely upon other people.

The West Ham Council reported that all their services had been diverted at the shallow depth near the Stratford Station. Then a miner put his pick into a forgotten 9 inch earthenware drain which served a row of small shops and houses. Before the pipe could be stoppered the air had blown out all the local drainage system, with considerable discomfort to a fishmonger washing his wares which were scattered to the roof and to an old lady *in seclusion*. Luckily nothing worse happened. The Transport Board had to provide a number of Elsan substitutes and claim on the Council. But even 5 lb per square inch can have an explosive effect. Years before a rival firm had driven a sewer under a low pressure a little too far with several feet of earth ahead. The face blew out and there was a fatality. This was borne in mind at Carpenter's Road Bridge.

Here there was only a 5 feet cover of untreatable ground, so after much ingenious work with bored piles to underpin the abutments, ground-water lowering as the water table was almost at ground level, and other expedients a steel piled cofferdam was driven, in which twin lengths of tunnel were built and buried in concrete. The top segments were made of cast steel to cope with the heavy live loads of the main road.

When the shields were at least fifteen feet away they were stopped and 6 inch horizontal boreholes were driven into the buried section, which had been connected to a pressure gauge. The air from the tunnel then filled the section until both were in balance before driving to join. The shields were then rebuilt on the other side to complete the drive.

COMPRESSOR PLANT

Owing to the potential hazards to the railway, duplicate electric supplies from two different sources were transformed down at each shaft, with duplicate rotary compressors, duplicate delivery pipes and cross-connections between every machine. In addition, two Petter diesel engines at each shaft could, by an ingenious device, start up within ten seconds of a current failure and take over the electric-driven compressors to which they were coupled. This duplication proved its usefulness.

As there would be considerable differences in pressure when the shields from shaft 1 junctioned with the tunnels at shaft 2, concrete headwalls, 5 feet thick were built at the west end of No 2 drive. I insisted upon building a 12 inch flanged pipe in each headwall, telling my deputy, who questioned it, that I did not have any reason except that it seemed a useful precaution.

The Transport Board then decided to carry out extra tunnelling for ventilation at shaft 1 so air had to be maintained at 24 lb for many months. So it was possible to connect one of the duplicate delivery pipes from the shed at shaft 2 and to pass pipes for high pressure air, electric cables and telephone through the other one and so save money by using one compressor shed for both drives.

AIR LOCKS

In major, large diameter sub-aqueous tunnels it has long been the practice to design and construct steel air-locks for men, materials and emergency, all built into steel prefabricated bulkheads, through which the many services are passed.

In early days air decks in vertical shafts were bolted to flat circular steel plates bolted to cast iron linings. In the 1920s we learned of an accident at Deptford Power Station which brought a change. Such an air deck burst at a pressure that was not very high, which dislodged a timber stage which fell on men in the bottom and several were killed by drowning. The investigation proved that the stress had caused bending in the cast iron flanges which broke off. Since then it became the practice to incorporate a steel ring among the cast iron rings to which the deck joists could be firmly connected.

For access passages or even in running tunnels of 12 feet diameter it was thought simpler and cheaper to build concrete mass walls 5 feet thick, which filled up the bosoms of the segments and to provide standard steel doors and frames, built into the concrete. In this case duplicate pipes were built in for all services with several extra in case of unforeseen need, as none could be added. This is possible with a steel bulkhead and judicious welding and burning.

FIGURE 46. *Twin air locks in 10 ft. access tunnel. (Mott, Hay & Anderson)*

At the east shaft, where air had not been anticipated, the design called for short cross connections to the running tunnels from a 15 feet diameter shaft sunk between them. When air proved necessary this would have entailed four air locks, one for each shield drive. Instead, at our own expense I drove an access passage between the tunnels and made the cross connections further from the shaft. In the access passage a steel plate vertical diaphragm suitably stiffened was built between the concrete headwalls, in which four doors were built, forming twin locks. This produced saving in time and money. **Figure 46**.

At shaft 2, I was instructed from *higher up* to build in a 6 feet diameter steel boiler lock, bought cheaply from a rival. This led to three educational incidents. When sitting in the lock as pressure neared 15 lb per square inch, we noticed that the rivetted seams were opening up. Luckily I had secreted a blacksmith-welder among my Black Gang who was able to weld up the seams before the gang came out.

The doors proved to be of cast iron. The inner one cracked so the miners had to be withdrawn and made to erect strong timber supports before locking out.

Measurements had been taken so a steel door was fabricated to fit. The miners then had to lock in and fit the door, before it would be possible to lock out. On enquiry it was found that the rival had ordered the lock for an escalator tunnel only needing 5 lb pressure and had designed it accordingly. Since then I have always held that any air lock at manufacture should be tested (hydraulically for safety) up to 75 lb pressure, even if this is beyond workable conditions. Air pressure released can be extremely explosive and human frailty incurable.

The third incident was caused by the defective design of this old lock. Air lock doors should be set so that they close under their own weight. In this case, at the end of a Saturday shift, a fitter — on locking out — filled up the lock in case anyone else had to come out. The inner door swung open and as no-one was inside the lock could not be used. Luckily a spare door was available so a brick wall was built a few feet outside the outer door, incorporating the spare door and so forming a small lock, through which entry could again be made. Here the installation of duplicate mains made it possible to feed this lock while the second pipe kept up pressure in the tunnel.

Thus the fancied saving by a senior engineer who, for all his great experience proved to be *pathological on the subject of money* led to considerable expense and risk. This strengthened my belief in *the Nelson touch* — failure to see the signal when instructions from seniors are obviously misguided. There was one last lesson in this case. I was constantly being told to remove every other bulb to save light in the tunnel, which I ignored. Later a loco driver was fatally injured when not looking where he was going. The Factory Inspector said to me, "I congratulate you on having such a well lit tunnel. Otherwise I might have had to take proceedings".

One other example of the Nelson Touch. I was also instructed to try to get the local Police Surgeon to test the men for compressed air working as Mr Jones had always found them cheap. I asked the Stratford Deputy Surveyor for the name of the best local doctor. His son took on the task at a much higher fee than had been hoped for. The first thing he did was to spot that one of our older experienced Pit-Bosses was a diabetic and prescribed accordingly. But as he was a valuable man the Doctor allowed him to carry on.

One Saturday afternoon, (we always worked on Saturdays) this man collapsed on his way home and was brought back and put in the medical lock as a case of bends. Dr Grogono came at once, took one look through the spy hold in the lock door and said "Bring him out at once. It is a diabetic collapse", and so saved his life.

It was not until the Second War that working conditions became a serious consideration. Good walking boards, good lighting and good ventilation are essential, but were often ignored by our predecessors.

RIVER INTAKE AT CANNON STREET

The following quotation is from *The Choice of Expedients* (Harding, 1946/7):

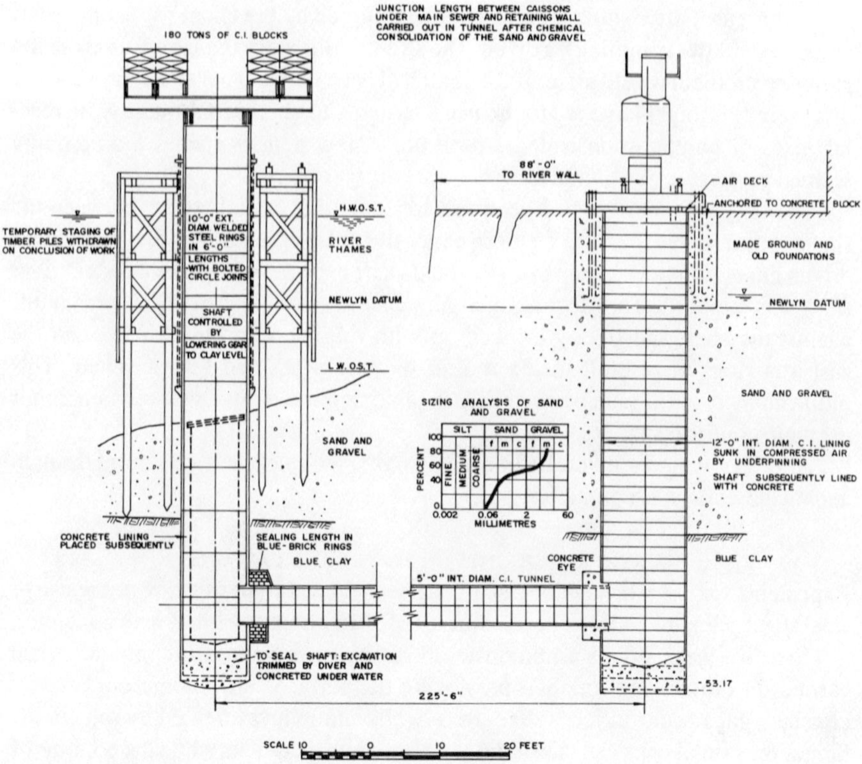

FIGURE 47. *Shafts and tunnel for river intake by Cannon Street bridge.*

In smaller problems it is sometimes sounder engineering to seal a cylinder or cofferdam by under-water concrete, than to pump out in risky circumstances.

" The paper shows an intake to provide river water for fire fighting. It was started just before the war for the London County Council. The land shaft under the proposed pumping station was sunk in compressed air through the gravel strata as compressed air was required in the tunnel. As heavy concrete foundations were required, the air-deck was placed at ground-level and the anchorage concrete formed part of the foundation. The 12 feet shaft was sunk by underpinning. The river shaft was sunk in free air from a temporary timber stage in the river. **Figure 47**. Steel rings were used, and screw lowering gear, until the shaft had entered into the ground. A kentledge frame and 180 tons of kentledge were used to force the shaft to a depth of 36 feet below river bed. It often happens that small blows can occur when a shaft penetrates the London Clay as a stone under the cutting edge may leave a small track through the clay sufficient for the outside head of water to cause erosion of an increasing nature.

" In this case, after some argument, it was agreed by the client to seal the shaft under-water after trimming by diver. The slight additional expense ensured that no pumping problem would arise. If the shaft had been pumped out and entry of water discovered, it might have led to the usual struggle to obtain continuously increasing sizes of pumps in an isolated position. A few pounds spent on a certainty seemed preferable to a gamble.

" The original intention had been to carry out a horizontal boring between the shafts, but this was found to be more expensive than a tunnel. A 5 feet tunnel was driven under compressed air from the land shaft until the outside of the steel river shaft was exposed. To effect a junction a chamber of larger diameter was built, against the shaft and tightly bricked up with ring work in blue bricks, and the cast-iron rings were built inside it after the brickwork had been grouted. This old-fashioned use of blue brick to help in obtaining water-tight work is sometimes useful in confined spaces.

" After the compressed air had been taken off, the steel shaft was burnt through and the junction was completed".

FURTHER THOUGHTS ON SHAFTS AND CAISSONS

Expedients for use when sinking shafts have advanced from freezing or the use of steel sheet piling to walls of bored piles or of slurry trenches filled with concrete.

There are weak points common to all methods of work below ground when carried on out of sight of the operator on the surface. Piles come out of their clutches, gaps occur in ice walls, slurry walls and in areas treated by injections. Sometimes small gaps can lead to large unforeseen operations while the bottoms of cofferdams often blow in spite of the best intentions.

When choosing such processes, the following points must be considered:

Is the ground suitable for their employment?
How can the seat of operations be reached?
What degree of success must be achieved?

Compressed air caissons have dwindled in popularity due to the above rivals and the need to pay higher wages and to neurosis over bone-necrosis. But there are few more certain methods of reaching the required destination safely than a well-designed caisson, which also simplifies the construction work within its protection. Experience has established a firm belief in the eventual economy of a policy of *belt and braces*. In the words of Mr Prentis, "No contractor ever went broke on the contract which he did not get".

Steel, though costly, is ideal for the shell of the working chamber. Mass concrete, when cast inside steel plating to act both as kentledge and as permanent wall, can compete with the cost of reinforced concrete with repeated setting and striking of shuttering. The walls must be integrated with the working chamber. On the Tyne Bridge it is recorded that a brick shaft sat on the steel working chamber which sank, leaving the brick shaft behind.

The modern Gowring vertical air lock with its twin man-locks is superior to continental designs (Storey Wilson & Sully 1949). The method of passing the crane bond through the cover of the man-lock, which then travels around above the cylindrical bottom-opening skip was an advance. The vertical muck-locks can be lengthened by inserting a length of flanged tube so that tunnel segments can be suspended to pass through the lock when this is serving a tunnel drive.

Much was learned from inspecting the work of others such as the compressed air caissons at each end of the Maas Tunnel. These contain the ventilation plant and from a logical transition between the immersed tunnel units and the cut-and-cover approaches.

The Pump House at Uskmouth Power Station was 164 feet by 110 feet and 80 feet deep. Its use on the sloping, very soft ground was fully justified (Storey Wilson & Sully 1952).

The floating out and sinking of large concrete caissons to complete the breakwater at Boulogne Harbour, visited in 1960, was a wise move.

The Queenhill Bridge carries the Ross Spur motorway over the River Severn and its flood plain. The Engineers designed the piers to be founded in 54 compressed air caissons of various sizes. **Figure 48**. The writer was able to advise and instruct the Contractor on how to perform and what precautions to take. He advised them to obtain Gowring vertical locks from Priestley of Gravesend. After these had been hydraulically tested they were filled with compressed air at the works. A day was spent with the Contractor's Agent passing in and out of the man-lock and working the muck-lock so that he could take charge on the site with authority and familiarity with their operation (Gibb & Tansley 1962).

BRAEHEAD POWER STATION: SCREEN CHAMBER

Years later this method of working under an air-deck inside steel sheet piling was used once more. At Braehead Power Station, beside the River Clyde at Renfrew, a screen chamber was to be built inside a cofferdam some way back from the river. No 5 Larssen piles had been driven 43 feet down to rock, forming an 80 feet square, divided into four compartments, 80 feet by 20 feet. Four boreholes showed 20 feet of sand and gravel overlying a varying thickess of *leathery* clay-silt and fine river silt with some boulder clay above the uneven rock surface. Ground-water level was 10 feet below the surface so there was a head of 36 feet at the toe of the piles.

At Dagenham, due to the methods used, it was possible to inspect the lower ends of Larssen piles which otherwise would have been out of sight. This was another useful lesson as it showed remarkable distortion in driving and a number of cases of displacement out of the clutches.

As at Dagenham, after trouble had arisen, I was sent down to take over, this time as a Works Director. I formed the gravest doubts that the cofferdam could be excavated safely in free air as the slightest distortion of the piles, especially among

FIGURE 48. *Queenhill Bridge — Main pier caissons alongside River Severn showing Gowring air-locks.*

the boulders, would allow an inrush of silt under the head of water, especially as the piles ended on rock. After much argument I persuaded the Engineers (Sir Alexander Gibb & Partners) that compressed air was vital.

To do this we dug down to ground-water level and then built air decks in each compartment a few feet high so that walings and struts for the rest of the excavation could be suspended from the air deck and lowered as required, as they could not be passed through the air locks.

Rolled steel joists 22 inches x 7 inches were welded to the steel piles at 16 inch centres resting on previously welded steel stools. Then longitudinal joists were laid beside the piles above the cross joists and held down to the deck by steel brackets again welded to the piles. This was to resist the uplift when the air was on. Precast concrete slabs laid on the bottom flanges formed shuttering and the joists were concreted in. **Figure 49**.

Two air shafts were fixed to plates on the deck in each compartment. The piles stood up as a precaution against floods for seven feet above the ground. The space inside the piles down to the air deck was filled with earth as an extra precaution against uplift and to provide a flat surface.

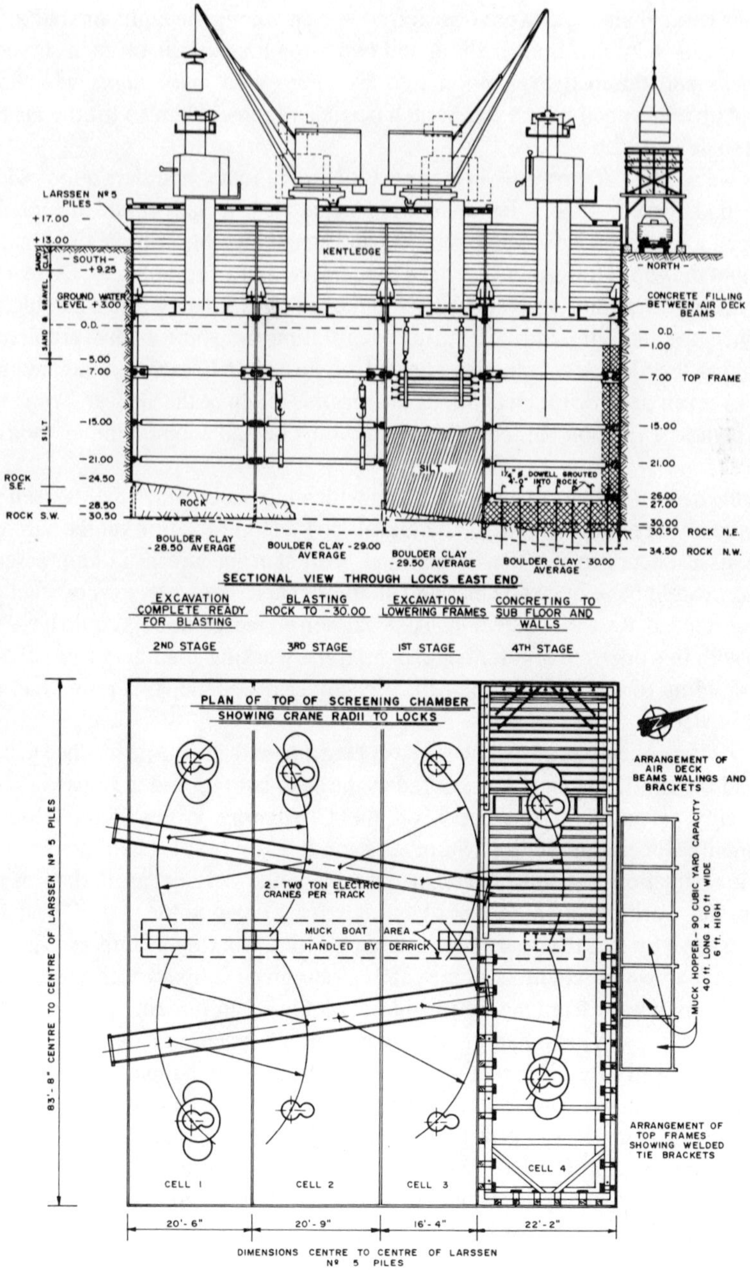

FIGURE 49. *Braehead Power Station — Method of construction of screen chamber.*

Four vertical air-locks were transferred in turn among the eight air shafts. Two crane tracks were laid on the filling and two 3-ton loco cranes on each served the air-locks and dumped excavation into six cubic yard *muck boats* which were picked up and tipped by a 7 ton Scotch derrick which could also lift the air locks from shaft to shaft.

As we worked downwards in compressed air we found boulders a foot wide or more had forced some piles apart. Then due to a misadventure the pressure dropped a few pounds for some minutes and a considerable volume of silt poured throught the gap. This proved the need for compressed air completely. On reaching rock the higher points were blasted off in the compressed air with no trouble.

When the late Dr Laurits Bjerrum was telling me some of his problems of designing the Oslo *Metro* in glacial silt, I suggested and described the method of making an air deck inside steel piling and that this might be the answer. Some years later he told a London audience of this and how he had adopted the method with success.

In the case of Braehead we were faced with a fait accompli as the Larssen piles had already been driven. As I have developed a slight fixation on the virtues of caissons it occurs to me that if one had been starting afresh a compressed air caisson would have been cheaper and simpler. There were 715 tons of steel piles and we needed 360 tons of steel joists. A caisson 80 feet by 80 feet could have been built with this or much less tonnage, even if the working chambers were divided. The building of the walls of the screen chamber inside the steel piling had been complicated.

In the type of caisson which we used at Dagenham the walls inside the steel skin formed the kentledge and were poured as the steel built up and the caisson sank. I especially recommend the paper on the Circulating Water Pump House at Uskmouth Generating Station (Storey Wilson & Sully 1952). I had carried out the site investigation for this and also descended the caisson as it did its most successful work, through 80 feet of soft silty clay down to the marl. The caisson was 164 feet by 110 feet with three interconnecting working chambers. The walls were built above the chamber as the caisson sank so no temporary cross strutting is needed which is far from the case inside a sheet piled cofferdam.

Mr W. Storey Wilson had great influence in his time on modern methods and some of us liked to tell him that we were his disciples, even though in a rival firm. He advised us on the caisson work at Dagenham which we had sub-let to him when he was with Dorman Long & Co. He and others wrote valuable papers for the Works Construction Division which might well be reprinted. This lively active Division was eventually emasculated as it stole all the potential audiences away from the main theoretical papers, but we tried to carry on its tradition of liveliness in the British Tunnelling Society meetings.

The Queenhill Bridge caissons have been mentioned. Sir Alexander Gibb &

Partners were the Engineers. There were many critics who claimed that steel pile cofferdams would have been better but I am convinced that the Engineers were right to stick to caissons. They were easily built on the ground, sunk smoothly and safely and acted as full size trial holes when decisions had to be made on when the Keuper marlstone which varied greatly from place to place was sound enough to found on. Fifty four caissons were a better choice than 54 cofferdams with all their internal strutting and pumping problems. (Gibb & Tansley 1962).

Chapter 10

ADVISING ON COMPRESSED AIR WORK (1956–1976)

Field Marshall Lord Alexander once said that "the success of a man in any profession will vary as the square of the number of people whom he knows". This is particularly true in the case of a single-handed independent advisory consultant. As with a doctor, every successful cure adds to his case-book of experience. Human behaviour can be as surprising as the behaviour of the ground.

In two cases when compressed air had to be resorted to unexpectedly, the behaviour of *the Engineer* in each provided interesting contrasts.

In the first case the contractor had undertaken to construct long lengths of concrete sewer pipes of varying diameters from 3 feet 3 inches to 3 feet 9 inches diameter under a provincial city. In order not to disturb the streets these had to be laid in timbered headings, driven from a number of manholes 360 feet apart. The ground-water table varied from 10 feet to 18 feet above the invert in different manholes. The intention was to follow the method used in another part of the city in gravels, of laying sub-drains leading back to pumps in each manhole. The contractor had qualified his tender to the effect that if compressed air became necessary he should be paid accordingly.

After work had started from several manholes at the north end, conditions were found to be worse than expected, although the adjoining work had been carried out in compressed air by another contractor. The faces proved to be in washed-down decomposed chalk known as *head*.

The state of the work had become so difficult that the writer was called in for his opinion by the contractor. After inspecting the headings the following is the conclusion of his report:

"The working conditions are such that it would be inadvisable in the extreme to attempt to carry on tunnelling as at present. The timbering in the heading and shafts has been carried out correctly with every proper precaution and no sign of negligence was visible. It is remarkable that the miners have persisted in their efforts. Of various expedients which might be considered to improve conditions, compressed air is the only one which gives assurance of safe and proper working".

My advice was to change to compressed air tunnelling, lined with concrete segments.

The City Engineer was reluctant to accept the extra cost so he consulted a long established firm of Consulting Engineers who supported my recommendation. After weeks lost in argument the writer's scheme was adopted.

Figure 50 shows the dimensions of the timber heading needed to install the concrete pipes, not forgetting the need to back fill and in places leave timber in. The outline of the 5 feet 6 inch rings shows that the excavation was no greater. The rings can be united and space given for flushers to inspect and clean out the sewer if needed.

The writer's previous firm lent a vertical air lock and air compressor. The capacity would be small so a diesel high pressure compressor could be used as a stand-by when equipped for changing to low pressure. The 10 feet I.D. shaft was sunk by grabbing. The joists which carried the steel plate needed to secure the air shaft were fitted into the bosom of the segments.

Some of the details were set out in a letter:

"The space between the underside of the joist flanges and the flanges of the concrete segments to be packed with dry concrete or packings to ensure a bedding and the shuttering for the air-deck which has already been erected can be suspended from these joists, so that the concrete air-deck can be cast as one. It is important that the up and down stresses of the air pressure and the counter-weight do not cause any bending in flanges.

"If the interior of the ring above and below the air-deck can be filled with concrete so much the better. The underside of the concrete air-deck can be washed with a cement wash when pressure is applied in order to stop leakages if this seems necessary. The space above the air-deck can be filled with water to counter-balance the air pressure, the weight of the segments to be ignored as a safety factor. Thus the weight of concrete plus seven tons of air lock requires enough additional dead weight to balance the pressure likely to be applied.

"Before the openings are made in the shaft, the sides of the opening should be supported by timber horse heads placed in such a way that a dummy ring can be built and a combined ring of concrete placed outside it to be worked into the flanges of the shaft, so that it can form a circular lintol, preferably carried up to the underside of the air deck to add additional support. Each eye should be done separately, and a few rings of tunnelling carried out at one end to make back-room before turning over to drive in the opposite direction".

This letter was to emphasise the need for care however small the work.

Having set the ball in motion the adviser retired from the scene as the firm were quite competent and driving then went quickly and smoothly. At the other end of the work, well-point work was feasible and successful, after the writer's advice had again been supported by another firm of consultants called in to criticise it.

The next surprise was to learn that the Engineer had stopped the compressed air drive just as it was about to pass in front of a factory. He then engaged Foraky, somewhat to their surprise, to freeze the ground for a length of 420 feet where the

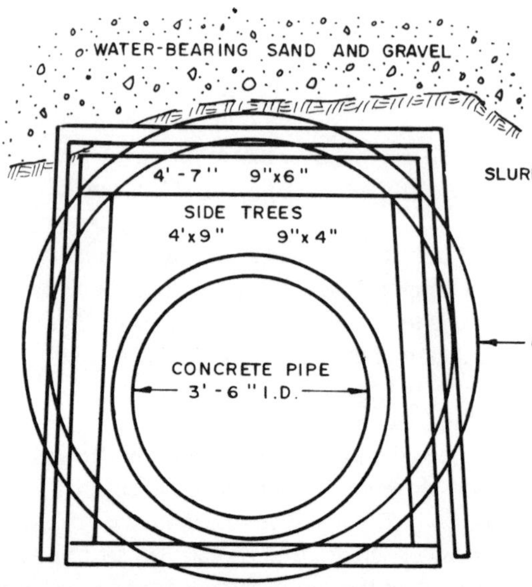

A

SECTION OF PIPE SEWER IN HEADING

B

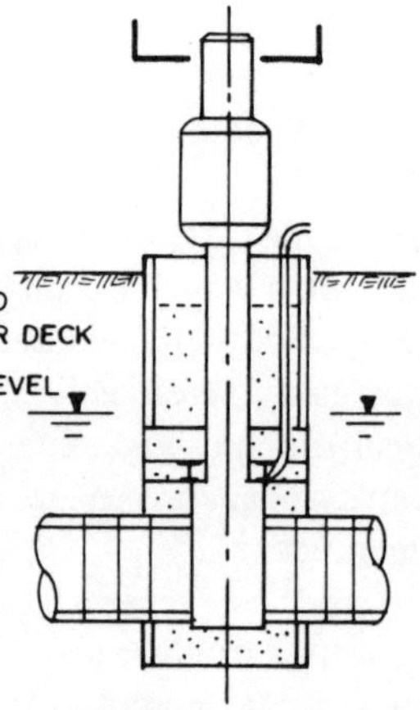

FIGURE 50. *Pipe sewer in heading, Norwich — Shaft with vertical air-lock.*

invert level was 20 feet below the road surface. Nearly 300 boreholes were needed for the freezing which took 3½ months and had to be maintained for some weeks more during the tunnel driving. The face had become in effect a rock tunnel, so hard was the frozen gravel and sand. Thus the work took several months longer than if the drive had carried on in air.

Why freezing was suddenly adopted remains a mystery unless it was an attempt to avoid further payment for compressed air work, though in the end the cost of freezing was not passed on to the contractor. The City Engineer may have been at the mercy of his committees who can at times interfere with the powers of the Engineer under a contract.

The following notes, dated 7 May 1963, made by the Contractor may have value:

"NOTES ON EXCAVATION IN FROZEN GROUND"
Sewer Tunnel at Norwich
Length 420 feet. Depth below ground under street: 19 feet.

To carry 33 inch diameter concrete pipes.

Cross section 5 feet wide, 5 feet 6 inches to crown. Top half arched 2 feet 4 inch radius. Only occasional temporary supports.

Two faces excavated, one from each of two manhole shafts.

Gang: 2 men in the face, excavating with compressed air hand tools.

Frozen sand and gravel with large flints.

Progress 1.12 feet forward per 12 hour shift, two shifts per day.

Area of face about 27 square feet, 2.48 cubic yards per lineal yard.

Time for excavating 420 feet with two faces: 4 months.

Men worked in spells of three hours in leather thigh boots and heavy protective clothing in 24 degree Fahrenheit of frost, – 13 degrees Centigrade.

Ground hard to dig, air picks made holes in face and ground broken out between them. Each man took a turn at the face while the other rested or changed tools. The air hammers froze up every twenty minutes. Spare hammers were stacked round a brazier on the surface and constantly changed. A filter on the air line reduced water content.

In a larger tunnel warm ventilation should be applied and tools heated in the tunnel: at Norwich the flints in the gravel made digging more difficult. Excavation about 0.9 cubic yards per shift by two men working 9 hours in every 12.

Area frozen about 11 feet square, cost £18,000 plus extra beyond three months after freezing. After setting up the plant twelve weeks taken to freeze the ground. Road considerably damaged by frost heave.

Delay at first as a laundry was pumping water from a nearby well and this caused a break in the ice wall.

NOTE: the work should have been done in compressed air as plant was on site. Reason for freezing — *local politics*".

AFTER EFFECTS OF FREEZING AND THAWING

A trunk sewer of 4 feet diameter concrete pipes had to cross under a Railway bank at Rainham, Essex. For this purpose a 6 feet diameter tunnel, lined with concrete segments had been driven below ground-water level for 90 feet between manholes under the bank, and the pipes laid within it. In order to drive the tunnel the ground was frozen by pipes driven at a rake from either side of the two rail tracks. **Figure 51**.

In due course, after the ground had thawed out the rail tracks sank 18 inches above the sewer and the concrete segments cracked. During discussions on alternative methods of reconstructing the sewer without damaging the tracks the cause of the settlement became an important factor. In order to eliminate several theories, tests were made by George Wimpey & Co. Central Laboratory on samples from boreholes inside and outside the frozen zone.

The ground, overlying the Thames gravels, which had been frozen, consisted of recent estuarine deposits of soft to firm organic silty clay with numerous pockets and layers of soft dark brown and black clayey peat. The conclusions in the Wimpey report carry a message:

"It is possible that the water held within the cellular structure of the peat and not naturally able to drain, expands on freezing and bursts the cells containing it. On thawing this hitherto captive water is free to drain and does so, producing the observed reduction in moisture content. At the same time, collapse of the cell walls and *packing down* of the soil causes an increase in bulk density.

The greater the density of the soil after thawing and lesser moisture content means that the soil becomes less compressible since less further volume change is possible on loading.

Natural peat, though highly compressible, has relatively high shear strength due to the tough fibrous nature of its skeletal structure. Breakdown of this structure is therefore likely to cause a reduction in strength, which phenomenon was observed in the present investigation."

HIGHBURY STREET TUNNEL, VANCOUVER

The second case where resort had to be taken to compressed air presented a contrast in scale and in human behaviour.

In 1960 a main drainage tunnel was being driven in Vancouver under a ridge composed of glacial deposits below Highbury Street for a length of 2½ miles.

The Consulting Engineers (Dolmage, Mason & Stewart Ltd.) and the Contractors (Northern Construction Company & J.W. Stewart Ltd.) had a long record in tunnelling, chiefly in rock and had gained wide attention by their work in blowing up the Ripple Rock, which was a peril to shipping in the area, from a series of tunnels mostly below sea level in the rock.

The tunnel was being driven from north and south portals with the help of second-hand horseshoe-shaped shields while the ground was supported by steel arch ribs at intervals with poling boards between. The ground rose 300 feet above the invert and was explored by a number of boreholes. Under the advice of Dr Peck

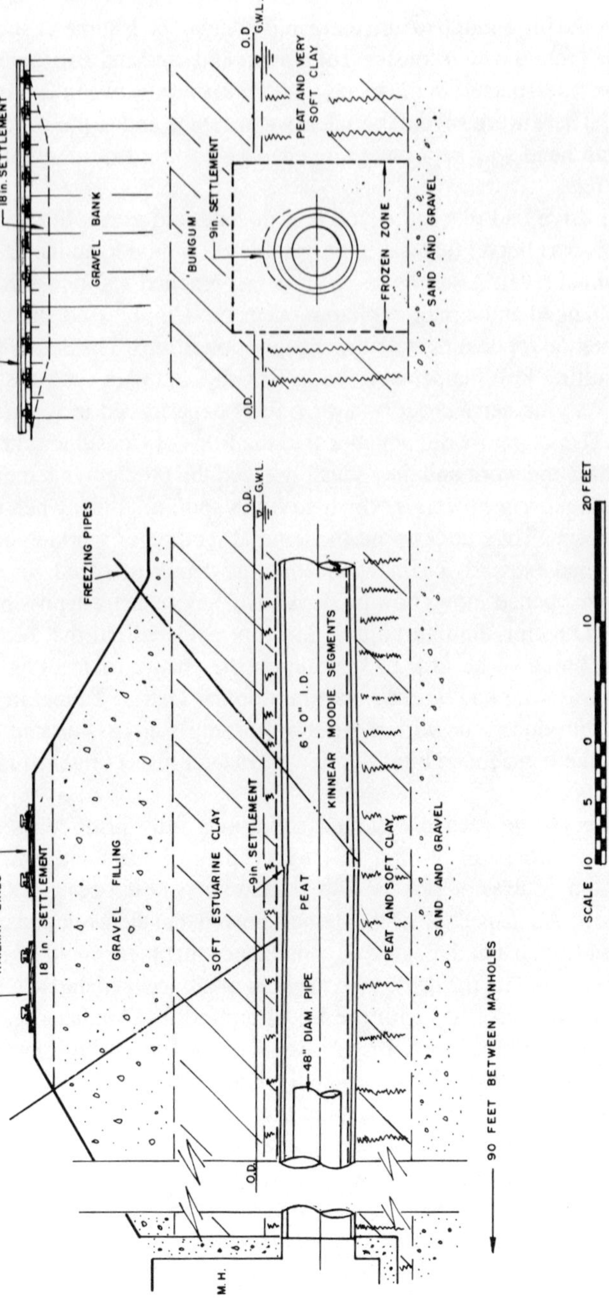

FIGURE 51. *Sewer crossing under railway after freezing.*

these were bored big enough to turn them into filter wells. **Figure 52** shows a rough section of the ground which consisted of gravels and sand and also unpumpable silt and fine sand interspersed in the usual random fashion between layers of till and glacial clay. There were several perched water tables and at the south end water under artesian head was kept below invert by pumping from the bored wells in Highbury Street.

The north drive had reached 6,700 feet in sand and gravel but without water trouble and it was hoped that the impermeable till would keep the ground-water above the tunnel level. After the south drive had reached 1,500 feet the ground in both faces changed and a major collapse occurred. Compressed air seemed to be the proper method for dealing with the changed conditions. The Senior Partner of a British consulting firm happened to be in British Columbia and insisted that the work could only be carried out by a firm fully experienced in working in compressed air. This certainly did not appeal to the firms of Consultants and Contractors engaged on the work and they sensibly faced the problem together instead of taking up *adversareal positions* which so often spoil relations when contractual difficulties occur. They understood the general method of working and felt that they only needed extra advice on the niceties and details involved.

What then happened shows how modern travel has made fresh possibilities. The Contractors' Director, finding himself in Montreal, realised that he was nearer London than home so he flew to London on the chance of meeting Sir Robert Wynne Edwards who had worked with him on the Detroit Tunnel in the 1930s, who in turn introduced the writer. After a morning's discussion and a visit to a travel agent, the latter found himself a few days later inspecting the tunnel face in December 1960.

As far as could be seen the ground conditions fully justified the change to compressed air working. Within this rocky part of the continent where the consultants and contractor practised there had been little occasion to tunnel in compressed air. An inspection of the tunnel showed that the work to date had been carried out with such a high degree of competence that there could be no doubt of their ability to complete the drive with the help of advice on details.

Then followed a meeting with the Local authorities, Medicos and others in which the virtues of the latest British Regulations for working in air were so emphasised that they were adopted officially in place of the much older ones which called for two short spells per shift rather than the modern emphasis on one long spell per shift controlled by decompression rates.

A long discussion was then held with the consultants and contractor in the form of a *teach-in* to discuss all the details which had to be explored, and which was recorded on tape for future reference. These covered the types, size and number of compressors, after-cooling, oil separation, standby plant, duplicate mains, instrumentation and so on.

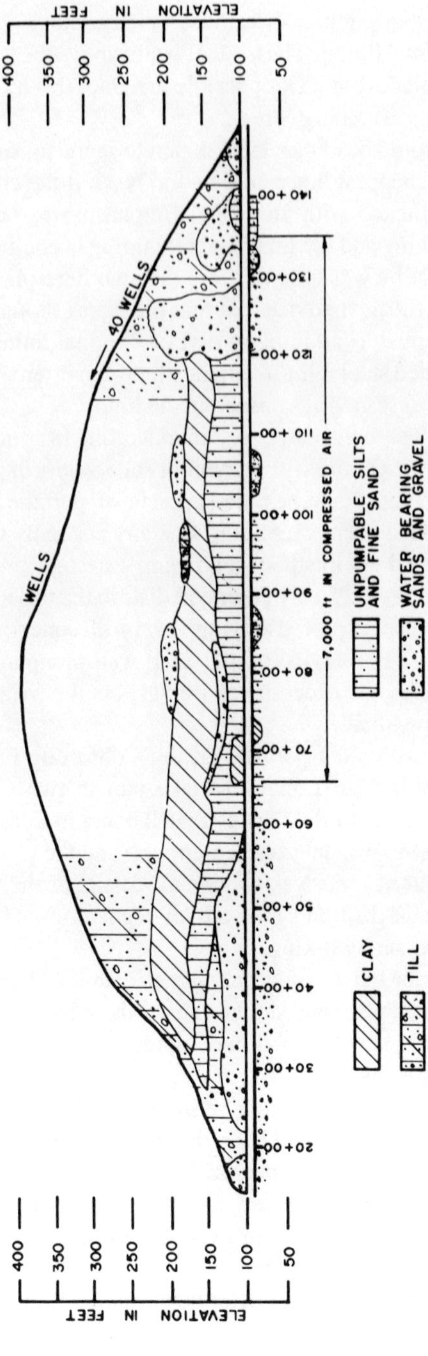

FIGURE 52. *Highbury Tunnel, Vancouver — Geological section.*

The writer advised that circular Greathead shields should be used with continuous primary segmental lining. The head of water above the tunnel was possibly as much as 100 feet in places but it was possible to reduce this by pumping from the bored wells strung along Highbury Street.

The primary lining could be of cast iron, concrete segments or steel liner plates. The last would be the cheapest but would be too weak if the compressed air was released unless strengthened with arch ribs. Emphasis was laid on the need to consider the primary lining and the final concrete lining in conjunction as a whole. The contractor said that he was able to press his own liner plates from Japanese steel. On this decision the next advice was that the plates should be grouted up in position and then accepted as an integral part of the final lining, and so act as a substitute for the intended steel reinforcement which can often obstruct the flow of pumped concrete and cause cavities unseen in the lining.

Two advantages outweigh the extra cost of concreting in compressed air. First, as the conditions made the possibility of a *blow* or sudden loss of pressure remote in the extreme, the air pressure could be relied upon to support the liner plates. Extra ribs would otherwise have been essential in free air. Secondly there would be no need to attempt to make the joints between plates watertight, whereas in free air without caulking, water would freely enter and disturb the unset concrete.

Once the details had been explored a flying visit by all concerned was paid to Dr Peck at Urbana, Illinois. His advice on the deep well pumping was invaluable, though he did not agree on the necessity of shields, but the writer held to his own opinion which proved justified.

A few weeks later two of the partners of the Consulting Engineers and two Contractor's Engineers flew to London to take part in two visits which I had arranged for them. One was to the Clyde Road Tunnel in Glasgow, then in full swing. The Engineers (Sir W. Halcrow & Partners) and the Contractors (Charles Brand & Sons) went out of their way to give all details of the working arrangements, for plant, air-lock drill and methods of regulating and dealing with the medical side of high pressure working.

This tunnel was shorter but considerably larger than the Highbury Tunnel and needed more sophisticated plant and shields, so another visit was paid to Belfast to inspect a 14 feet diameter sewer tunnel in compressed air, using a simpler shield and bolted concrete segments.

By then the writer had located two hooded shields used by John Mowlem & Co. for cooling-water tunnels, which had been driven for only 500 feet. These were obtained second-hand, through their makers, Priestleys of Gravesend, who reconditioned them, painted on erection markings and dispatched them to Vancouver. In addition Mowlems were able to lend an experienced foreman for six months as there was a lull in their tunnelling work.

While awaiting the shields the Contractor had installed appropriate compressor plant at each portal and had lined part of the completed drive. Concrete bulkheads had been built as more appropriate than steel and air-lock doors had been designed big enough to allow the passage of Moran rotating concrete transporters mounted upon track.

The shields were 11 feet 6 inches O.D. so the liner plates were designed to 11 feet 3 inches O.D., 18 inches long with punched bolt holes, ten segments to a ring and 2¼ inches deep. The shields had been used for 10 feet 6 inch internal diameter cast iron lining and had been made to pass in sections through airlock doors.

In order not to incur expense and loss of time in altering the available shields to a slightly larger diameter, my argument was accepted that a few inches less theoretical cross section would not discourage the passage of the effluent as most hydraulic formulae contain a constant K which is far from exact.

Figure 53 shows the original section and the final one. The original area of excavated face was 134 square feet. With the circular shield the area was reduced to 104 square feet.

With the arrival of the shields and the foreman, work proceeded to a successful conclusion and all subsequent advice was carried out by *correspondence course*. Clay pocketing was used in gravels. The fine silt at times was described as unpumpable and pressures ran up from 20 lb to 34 lb per square inch. This sometimes raised the level of the water table and was reduced with better effect.

On August 25th, 1961, Mr Lambert, the contractor's director, wrote: "We are now at 29 lb at the south end and 15 lb at the north, and there does not seem to be much rhyme or reason for the variations of the ground which we are encountering. The face can change completely in a single push and of course with the natural ground-water level so high above us, we keep our fingers crossed and view with concern any increase in pressure.

"We have been very fortunate I think, generally speaking, with the incidence of bends, which has been quite low, considering the inexperienced crews we have had to use and the fact that we had to start green men under such pressures".

On October 6th, 1961 he wrote: "I might say that our air consumption has been very low. When we first started at the south end it went up to about 1,200 feet a minute until we got away from the ground which we had disrupted to the surface. Since then it has only been three to four hundred feet a minute. When we raised the pressure to 34 lb it went up to 1,500 feet but when we stopped and bulkheaded the face, the consumption has dropped back to normal. I might say that at the north end we are carrying at about 22 lb. It has a much more homogeneous condition than we are at present facing at the south end …

"Incidentally, you speak of Dr Peck's wells. They have been, of course, a life-saver, as we have been pumping well over 1,000 gallons a minute at the south end and it would be impossible to operate without them".

The British foreman left for home earlier than expected but by then routine had

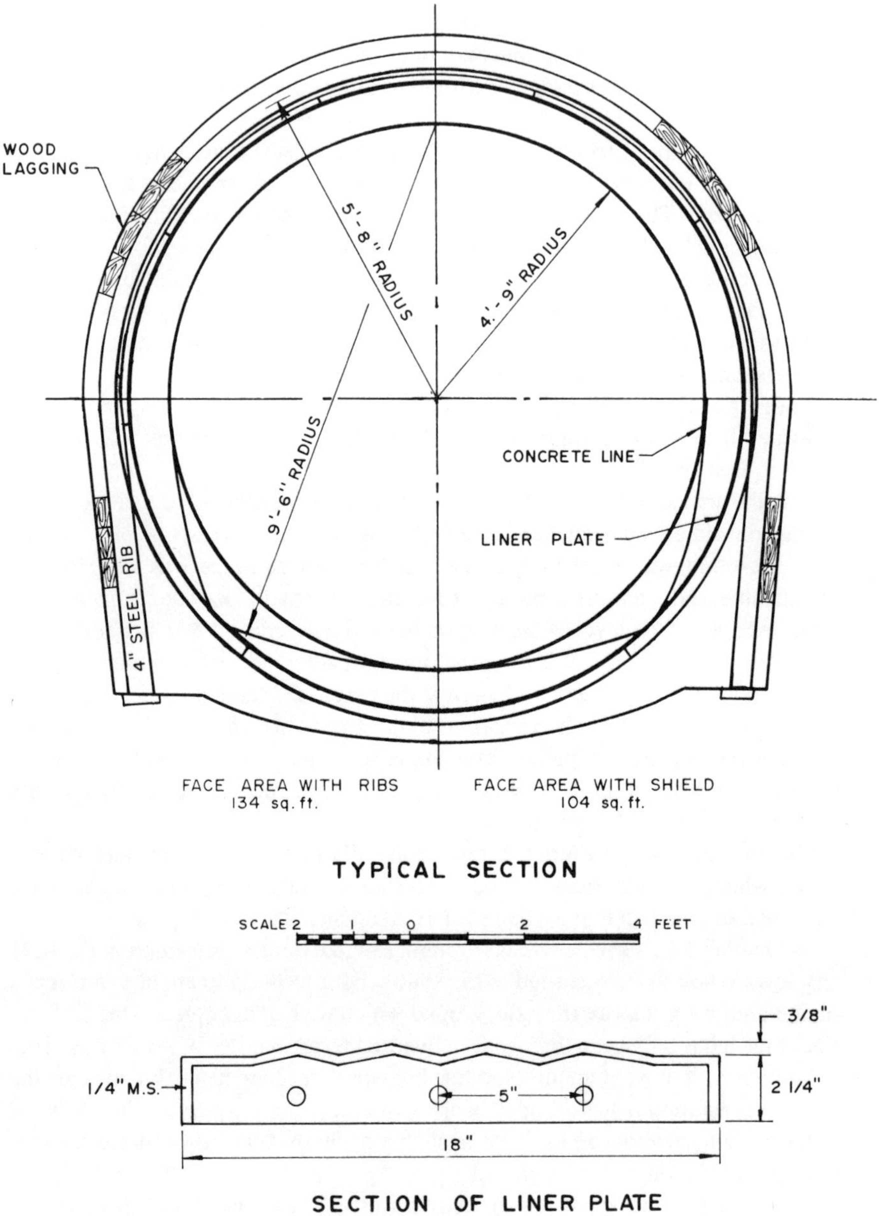

FIGURE 53. *Vancouver — Sections of tunnel and liner plates.*

been established. In one reply I wrote: "Do not let the men get over-confident now that they are on their own. What gives me the greatest pleasure in the Vancouver Affair is the proof that you and your firm are perfectly capable of carrying out work so long as you have been given *the rules of the game*, as distinct from our friend who said that it must be done by an experienced compressed air contractor".

During the Vancouver incident in 1961 diaries show that fourteen other assignments were woven in, of which only three were concerned with tunnels — a trip to Sweden over rock tunnels and to Spain, and one to Rome to advise on how to tunnel under the River Tiber in compressed air (which did not happen).

Morrison-Knudsen had an interest in the Northern Construction Co. and were aware of the work described above and this led to another episode.

RAPID TRANSIT, MISSION STREET, SAN FRANCISCO, 1966

The Work of the Channel Tunnel Study Group, 1958–60 (Bruckshaw et al 1961) is discussed under Site Investigation. But it also brought contacts with U.S.A. firms who had been asked by Technical Studies Inc. to provide reports. These included Morrison-Knudsen, Bechtel Corporation, Brown & Root as well as the consulting engineers Parsons, Brinkerhoff, Quade & Douglas.

In November 1966 Mr Jack Bonny, President of Morrison-Knudsen invited me by transatlantic telephone to spend a week in San Francisco from 12th — 18th November on part of the B.A.R.T. scheme (Bay Area Rapid Transit). This was to consist of twin tunnels 5,200 feet long, lined with weld-fabricated steel segments 2 feet 6 inches long of 18 feet O.D. and 16 feet I.D.

For the whole week the top management of the Contractor sat in almost continuous session to discuss every aspect of the work before starting; together with a number of advisers or sub-contractors on appropriate problems.

The Engineers for the whole B.A.R.T. project were Parsons, Brinkerhoff, Tudor-Bechtel. The specification which was clear and comprehensive called for the tunnels to be driven in compressed air by hooded shields but also with a system of ground-water lowering by deep wells bored in the streets in order to reduce air pressures to 12 lb per square inch. Clause 7.6.4: "Tunnelling machines, if used, shall be of such a type as will afford adequate protection against the loss of ground, as specified, and will provide for the erection of lining within the machine".

Five sheets of borehole logs recorded the results of two, slightly conflicting site investigations, while a table of soil classifications described fifteen different materials. These ranged in degree from gravel through sand to silts and clay.

My first step was to transfer these logs onto the longitudinal section of the tunnel to realise their true significance. Some patches would be almost undrainable and perched water tables would occur in places. The borehole samples were inspected and it was possible in a report to relate different grain sizes to the results of other ground-water lowering works in England. Meetings were also held with specialist consultants and sub-contractors anxious for the work of ground-water lowering.

The working site was to be on a 2,100 feet radius curve which carried the drive under Market Street to follow under Mission Street until the next station at the intersection with 15th Street. At this centre of the contract the tunnels were to be driven in both directions from a permanent ventilation and pump shaft 95 feet deep.

This shaft was to be rectangular in plan, 62 by 28 feet overall. The walls were to be four feet thick of reinforced concrete but built in trench by the Icos method. The ground-water was to be lowered to fifteen feet below the formation level of the shaft by 20 or more deep wells. This was so that short lengths of tunnel in free air could be driven before starting the compressed air shield drives. This would entail four air locks for four shields, or if two tunnelling machines were used they would have to be dismantled and rebuilt for each half of the drive.

After studying the problem I concluded that the ground-water could not be lowered below the formation level due to the ground and that a drastic change of method would save much time, much money, and many wells. In my report, which I dictated at the time: —

"Alternative Method of Carrying out the Contract

"During our discussion I made the observation that the engineers had designed it *the hard way*, as it would be much simpler to work from the 15th Street end. This was a passing observation. But subsequently Mr Anderson, Mr Maclean and others have discussed the matter as a serious possibility and I will make my own comments.

"The contract design with the shaft in the centre of four drives means that great complexities occur right at the start. These are:

The sinking of the Icos shaft, plus the setting time of the concrete, controls the whole operation.

Numerous deep wells have to be sunk around the shaft and their success determined.

Large groups of wells are needed around the building at Gough Street and at Otis Street as these areas are the first under which tunnelling has to be carried out on the curve.

Complex work at considerable depth in starting the tunnels towards 15th Street.

The boring of back-shunt tunnels and the problem of eating them out when driving to 8th Street.

The installation of air-locks and changing them over after the first pair of tunnels are completed.

"If it is possible to start at 15th Street, tunnels are at a quite shallow depth. The initial lengths could be carried out in sheet piling as there are impermeable beds at that point and even then a steel pile cofferdam would be simple. It would be possible to build machines and air locks in the open and also the plan shows parking lots adjacent to this position. Even if the initial work had to be carried out from a shaft, the pressures would be low and the work simpler than at the contract working site.

"It is obviously more economic to use a tunnelling machine for long continuous drives with gangs working in a routine than to drive for half the distance, retrieve the machine and start all over again at another place. In addition, while the tunnels are driving from 15th Street, the Icos shaft can be sunk without it having to prove the key to the whole progress and it should be in position before the machines arrive.

"The dumpling need not be excavated until the tunnels are passed through and the essential problem of lowering the water below the bottom of the shaft can be dealt with at leisure. If the tunnels are driven through the shaft using temporary steel lining the dumpling can be excavated as far as is safe and a concrete air-deck placed and loaded with spoil. It will then be possible to excavate and concrete the invert of the shaft in air bringing in reinforcement and concrete from 15th Street and eventually clean out the shaft and temporary lining when the compressed air is taken off.

"But there is one great advantage. Wells are needed around the Gough Street and Otis Street areas and their influence may be sufficient with only two or three additional wells to lower the pressure in the shaft to 12 lb. This means that the attempt to lower the water to 15 feet below the shaft becomes unnecessary and possibly as many as 20 wells could be saved. This is a most important factor in the revised scheme".

After I left, my client passed my suggestion to the Consulting Engineers who welcomed it. So the working site was transferred to the shallow end at 15th Street. This saved over 20 wells and also gave time for underpinning a large Bank near the Vent shaft before tunnelling approached it.

Other advice was that the ground would stand up well for a soft ground tunnelling machine. "A highly sophisticated machine is not called for, but one which is basically a shield so that at any time the cutters could be removed leaving a shield as a fully working proposition". In the end they chose a sophisticated Memco Machine.

COOLING WATER TUNNELS FOR ATOMIC POWER STATIONS

Cooling water tunnels for intake and outfall works were an essential factor in the successive post-war programmes for Atomic Power Stations. Several Consortia of Manufacturers, Consultants and Contractors were formed to submit competitive designs with offers to construct each successive station. Additional advice was requested from several sources involving discussions on design, tenders and later, on problems of construction.

TUNNEL DRIVING AT SIZEWELL

Sizewell Power Station is one particular example of three involvements. The first was a visit to four potential sites on the Suffolk coast with consultants Lewis & Duvivier who were assisting in the choice of site from aspects of coast erosion, sea bed activity, past and future, together with their relation to cooling water problems. This led to the choice of Sizewell. (Gammon & Pedgruft 1962).

Then followed a request to help with the design and method of construction of the cooling water system being worked out by the consultants C.S. Allott & Son and the contractors, Taylor Woodrow Construction Ltd. This was in spite of a previous unsuccessful bid with the same group for Dungeness A, where conditions had been much harsher and so invoked a cautious approach.

A third involvement was a request to report to the English Electric Company "on special features of the Circulating Water System" with reference to possible after effects.

The Consultant-Contractor Group had driven the cooling water tunnel at Hinkley Point in rock into a large and ingeniously contrived caisson in the sea. They now wished to discuss more specialised details of work in compressed air, first at Dungeness and then as their experience was limited for the successful bid for Sizewell. Many pages of letters were written to confirm the details covering methods, air locks, air consumption and capacity of plant, hooded shields, type of lining, caulking, pumping out of sand and much else, both optimistic and cautionary.

Six marine borings had been carried out from a tower made of military trestling of square steel columns, erected in Lowestoft and carried to the site by a small crane ship. It rested on the sea bed during drilling and was re-placed on each site by the crane ship.

The borehole logs for the tunnel area showed "medium to fine sand, very dense (Norwich Crag)" for most of the length with layers of clay above topped with sands and gravel.

The intake was to be 1,800 feet long and the outfall 985 feet long. The most important factor was the request by the Central Electricity Generating Board that the shafts should be jacked up from the tunnel through the sea bed as pioneered successfully at Dungeness by Sir Robert McAlpinc & Sons.

The tunnels were finalised at 11 feet I.D., 11 feet 9¼ inches O.D. lined with unmachined cast iron rings two feet long. The cost of machining was considered too high, in spite of the writer's arguments that the saving would be cancelled out by the cost of caulking unmachined iron, which proved justified.

From the sizing of the samples the advice given was to allow for 10 cubic feet per minute per square foot of face, assuming that the caulking was kept close to the face. Electric compressors of 3,000 cubic feet per minute were installed with 100 per cent standby of diesel driven compressors in a single shed.

The driving proceeded safely with only a few points to record. Air consumption for some of the length rose to 25 cubic feet per minute. The face on examination was found to be made of separate thin layers of fine and coarse sand. Also the sand grains of the Pliocene Norwich Crag deposits were found to be lightly cemented and held apart rather than densely compacted. These had been broken and churned up in the borings and so gave the impression of a dense mixture of both fine and medium sand. The caulking had also fallen behind the shield due to the time needed to cut out wood packings and to the use of more material than would have been the case with machined iron.

It has often been necessary to impress on clients the fact that compressed air can cut through cement grout before it has fully set even in the case of machined joints. This can make a narrow *slice* for the circumference of the tunnel. If caulking falls behind, these slices, when multiplied by the number of rings can add up to a considerable area. It is good practice to lay a thin strand of caulking material in the caulking groove before grouting. This can be fully caulked when the caulkers follow behind, This can be more useful than a second grouting which can suffer in the same way as the first if the caulking grooves are not filled.

Extra compressors were easily installed owing to my insistance on duplicate air mains and valve connections. Also the compact nature of the ground allowed progress to be much faster than anticipated. The invert level was −49.50 O.D., the sea bed about −21 O.D. and the tide range was from +3.70 O.D. to −3.90 O.D. The pressure used was about 18 lb per square inch and only varied by itself for one and a half pounds with the tide. This surprised the contractor's young engineer but confirmed the writer's own experience. (Harding 1948).

The sand was pumped out of the tunnel through pipes built into the bulkheads which simplified track laying and cleaning out. Two means of closing the pipeline were insisted upon to prevent air from escaping through the pipe when not in use in case of careless action.

Shaft Jacking

The decision was made to jack up two shafts of 6 feet 8 inches diameter from each tunnel instead of enlarging the tunnel to accommodate a single, larger shaft. **Figure 54**. My advice, acted upon, was to drive the last part of the tunnel with fabricated steel lining. This contained a number of special rings with segments so designed that they would, in conjunction, provide openings for the shafts and also form roofs for the shafts when bolted together on to them.

Following the example of our early engineers, on my advice a rehearsal was staged in one of the inland sand dunes to make sure that all systems worked and that the drill was thoroughly understood. Six 100 ton jacks worked off a base laid in the invert and retracted to allow each ring of the shaft to be built. Emphasis was laid on the need to provide anchorages in case the friction became so low that the air

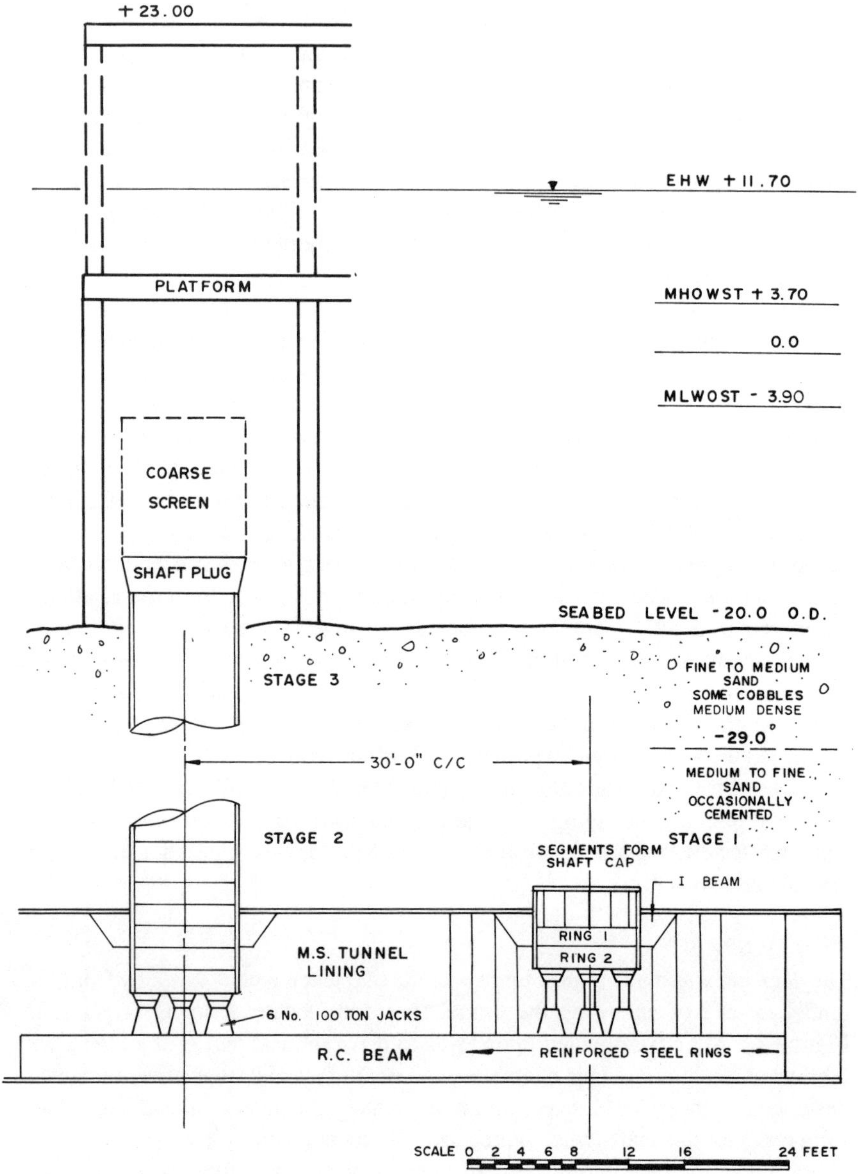

FIGURE 54. *Sizewell cooling water tunnels — Shaft raising.*

pressure could try to blow the shaft out of the tunnel. Holes were provided in case jetting was needed to overcome friction. The writer's prescription for successful jetting is to insist upon at least 100 lb per square inch pressure and at least 300 gallons per minute in combination.

In order to place screens and provide permanent platforms over the shafts, light stages were floated out and secured by driving piles connected to the legs. If large shafts had to be sunk at sea into which tunnels in compressed air had to be driven, the cost of such platforms would be considerably higher as they would have to accommodate plant, men's quarters, lowering gear and all the additional complexities needed for such work.

Finally the conclusions in my report to the English Electric Company (before the work started) were: —

"The upcast method for the construction of the shafts is novel, but should be economical and safe.

The associated permanent maintenance structure over the heads of the shafts can be of lighter construction than if the shafts had had to be sunk, and should be stable against sea damage.

These constructions are sufficiently small not to have any marked effect on the regime of the coast line and as the bed is comparatively stable the danger of silting is much reduced.

The risks of navigational obstruction are covered if the rules of the Ministry of Transport are fulfilled".

After twenty years, these conclusions have not so far been proved false.

THE PARIS METRO

The cases so far described merge from the obtaining of experience into the Age of Advising. The familiar contract condition that "only firms experienced in this class of work will be considered" can be circumvented by following the dictum of Sir Henry Japp: — "Hire someone who knows how". For intelligent civil engineers who have had some form of basic experience can undertake anything new, but advice can help. Sometimes a firm is diversifying for the first time into a new field, sometimes those with experience merely want reassurance.

This was the case with the old-established French firm of Fougerolle with whom contact had been established during the first phase of the Channel Tunnel Study. In 1962 they wished to discuss aspects of their tender for the new Metro extension from L'Arc de Triomphe, northwards to the River Seine. This was to be in medium to fine sand under slight artesian head, but with a thick bed of soft limestone just above the crown. If the tunnel could have been raised so that the upper part was in the solid ground much trouble might have been saved. This was not done as the designers wanted to use the down grade on leaving and up grade on entering stations.

The Metro Authorities were pressing Fougerolle to adopt the untried shield designed by the Metro engineers, with French logic. The intention was to achieve the age-old desire to contain the compressed air only in the face behind a diaphragm. Jacks to support the face boards were to be remotely controlled from outside the diaphragm while the miners scratched away and loaded up the sand by articulated tools, while they remained in the free air.

The personal advice given was prompt and blunt. Have nothing to do with it. The tender was unsuccessful but the client later expressed gratitude. On being asked why he replied that the shield had proved a complete failure. On a later visit, to the Metro being driven at Defense, the shield was seen to have been completely altered. Airlocks had been welded to the diaphragm and miners were working in compressed air, in safe conditons in such a small space, as the whole of the sand had been solidified chemically by pipes driven in advance from the surface, which must have been a costly operation.

The hope to continue the drive under the Seine with the original design was abandoned and the crossing completed by a succession of compressed air caissons, as had been done many years before.

PROPOSED COMPRESSED AIR TUNNEL UNDER THE RIVER RHINE, DUSSELDORF

In 1957 the well-known West German firm of Dyckerhoff & Widman asked for my advice on problems of tunnelling under the river Rhine in compressed air at Dusseldorf and on the use of their new design of a post-stressed concrete lining. As I was unaware of how much they knew I made a list before I set out of every small point which might need discussion if starting from zero. These numbered 160 items.

Points for Consideration and Decision

Preparations:
- Survey across river
- Lay-out of working site
- Choice of handling and winding plant
- Type of shafts or caissons
- Air locks
- Provision for lowering large sections before horizontal locks are built
- Offices, stores workshops, sheds
- Mess rooms, drying rooms, First Aid
- Medical air locks and medical care
- Power, heat, light, water
- Compressor plant, low pressure
- Compressor plant, high pressure
- Cooling water

	Battery charging
	Type of haulage
	Stand-by plant, pumping plant
	Low voltage current in tunnel
	Ventilation
	Air mains
	Provision for survey lines to tunnel
Shields:	Dimensions
	Type — hooded or plain
	Bolted or welded
	Steel or cast iron in parts
	Length of hood
	Length of tail (depending on segments)
	Number and types of jack
	Water traps
	Sliding platforms
	Muck chutes
	Face jacks
	Method of timbering and working
	Length of housing ring for jacks
	Travel of jacks
	Possible dismantling method
	Hydraulics and hydraulic pumps
	Trailing platform
	Possible *Jumbo* stages
	Method of loading away excavation
	Conveyors
	Erectors — on shield or on Jumbo
	Turning round of long segments
	Cleaning of inverts before erection
	Centering of segments for position
	Adjustment for *square* and *look-up*
	Method of bringing segments to face
	Removal of empty trucks
	Type of track to be used
	Type of crossings
	Propulsion of Jumbo
Segments:	Accuracy of casting
	Length
	Variation in bolt holes
	Provision of holes for carrying pipes, cables and stages

Whether bottom segments to be flat
Whether drain holes in bottom segments
Prevention of rolling
Ability to build with flush face
Choice of packings for correction
Taper of key
Grouting plant and method of grouting
Caulking, method and material
Second line of defence against water
Type of locos, skips and flat trucks
Type of sleepers if invert not flat

Air Locks: Horizontal air locks-type
Diaphragm walls — steel or concrete
Double walls or single boiler lock
Number of pipes through lock, spares
Duplicate low pressure
Duplicate high pressure
Telephone, water, power, light
Spare telephone
Pressure gauges, automatic recorder
Type of air lock doors
Material locks
Relative position of locks
Feed of lock off main at start
Taking off air during building locks
Precautions necessary for this
Drainage of tunnel
Sumps
Snorer pipes through walls
Whether to use safety curtains

Method of Working: Type of shaft — square best for launching shield if ground suitable
Possibility of air chamber in steel shaft
Two vertical locks
Use of decanting lock on surface
Upper air deck if shield to start in shaft
Method of entry of segments and possibly parts for tunnel air locks
Alternative shield chamber by hand
Method of winding
Disposal of excavation
Method of shield erection

Check for shape and accuracy of build
Breaking-out of shaft
Steel plates to be removable or cut out
Precautions if in circular shaft
Distance driven to allow for air locks
Shunting distance outside air locks
Whether clay pocketing necessary
Method of timbering face
Boxing up of face
Shifts to be worked
Erection of segments
Grouting of rings
Allowance for air loss at joints
Checking of shield for centre, square and overhang
Uphill driving and drainage
Whether air mains to face for ventilation
Extraction of oil from mains
Fire precautions — no smoking

Compressor Plant: Types of air compressors
Number or units
Varying capacity
If electric then diesel stand-by
Separate electric power supplies
Transformers — Bus bars, switches
After-coolers
Oil separation
Interchangeability of machines
Inter-connection of air mains
Telephone from shed to face
Need for Mates to compressor Drivers
Duties of electricians
Silencing of machines
Cooling water supplies by gravity
Cooling by pumping round
After-cooling, or water to waste
Secure foundations
Fireproof sheds
High pressure compressors
Receivers — size of mains

Surveying and Setting Out: Triangulation by day or night
Base line — length and location
Correction for temperature and curvature of earth

Type of instruments
Method of levelling across river
Beacons and lighted screens
Testing of tapes
Method of carrying lines and levels into tunnel through locks and down shaft
Setting of square marks, lines and levels
Recording of shield *shoves* and checks
Tables of offsets for curves
Keeping of progress plans and strata
Supervising staff

Sanitation

This led to a number of visits to their office in Munich and to Dusseldorf with many interesting consultations. The concrete lining which they had designed using their Dywidag patents was so arranged that a number of subsequent rings could be post-stressed with each other horizontally as well as the circular stressing of the segments. Trial erections of rings of 4.50 m were on show as well as tests on a plastic model in a powerful testing machine. I wrote a report supporting its use while stressing the importance of making the circular joints watertight.

The Dusseldorf City Engineer had asked competitors for this water supply tunnel to show that they had expert advice on the tunnelling. One firm who were not successful had in tow an American engineer of experience from a contractor and he pressed strongly for a cast iron lining. The size of the proposed tunnel had by now been reduced to about 3.80 m I.D. This persuaded the City Engineer to decree that the two-thirds under the river should be in cast iron and the remainder under the land with Dywidag concrete segments. In the discussions he also wanted me to take some overall supervision but this did not quite suit my individual practice or my obligations to my client.

The next move was the acceptance of a combination of Dyckerhoff & Widman with Hochtief who had engaged Kinnear Moody to help them and to provide expert help and foremen. So in due course the tunnel was successfully built by a highly competent group without my further appearance.

A few years later I had more pleasant and friendly discussions with Dyckerhoff & Widman over details of their estimate for the new Munich Underground with its ambitious designs for the stations. This included a visit to a heading in Dachau some miles out where the nearest outcrop of the sub-Munich clay was exposed in a heading. In this case they were not the lowest competitor.

About this time there was an upsurge of interest in city *Metros* on the continent and as tunnelling in soft ground, especially in compressed air, had been scarce there for many years, a number of firms were searching for advisers among British consultants and contractors. Thus I found myself in Rome looking into the problem of tunnelling in compressed air under the River Tiber for the proposed underground railway, on behalf of Farsura of Milan. I no longer have my papers and reports but as far as I remember there were no insuperable problems. One warning was to investigate beforehand for a possible buried channel. But the whole project was shelved.

Then in the June/July number of *Tunnels & Tunnelling*, 1979, there appeared an article on the Rome Metro. The problem of tunnelling under the Tiber has been solved by the tunnels suddenly coming to the surface to cross the river by a bridge. This is so up-to-date that it is protected from terrorists by bullet-proof glass. It is safer down below. The article then describes the highly successful waterproof lining by using rings of 5.50 m diameter which are pre-stressed horizontally and on the circle by Dywidag bars. This, the writer claimed, is a World First. But what about Dusseldorf twenty years before?

MEXICO CITY: MAIN DRAINAGE

On a visit to Sweden to inspect Widmark & Platzer's rock tunnels and ladder drilling, I met Senor Suarez, a Mexican contractor who was working with Widmark & Platzer on rock tunnels in Mexico. As a result, in 1962, both firms invited me to discuss preliminary details in Mexico City over the proposed new drainage scheme.

Mexico City is built on reclamations in an area of lakes contained within a ring of mountains. The subsoil for a depth of 200 feet contains thick layers of highly colloidal clay with water contents up to 300 per cent of the dry weight. The intention was to deal with the vastly increased size of the city by an extensive tunnel system, possibly going all the way to the mountain range 30 miles distant.

I took with me a copy of the I.C.E. paper *The Drainage of the Valley of Mexico* (Body 1900), Lord Cowdray's chief man in Mexico (who had married my aunt). This described the large canal which carried all the drainage to a tunnel through the distant mountains. Discussion took place with a Minister on the generalised choice between carrying the considerable flow by gravity in tunnel to the mountains or to establish pumping stations outside the city and lift the flow to a larger canal. The tunnel would involve initial heavy capital expenditure whereas pumping would involve continuous heavy annual cost with undesirable risks in the case of mechanical or electrical failure. **Figure 55**.

As no detailed scheme had yet evolved, discussion followed on some of the possible 162 items in compressed air tunnelling and the possible behaviour of the unique soil. The report to Widmark & Platzer discussed the canal, future schemes, soil conditions, future site investigations, tunnelling procedure, working shafts, connections to existing sewers, working sites, tunnel linings and conclusions.

Dr Terzaghi had written that, "the most spectacular regional subsidence is going

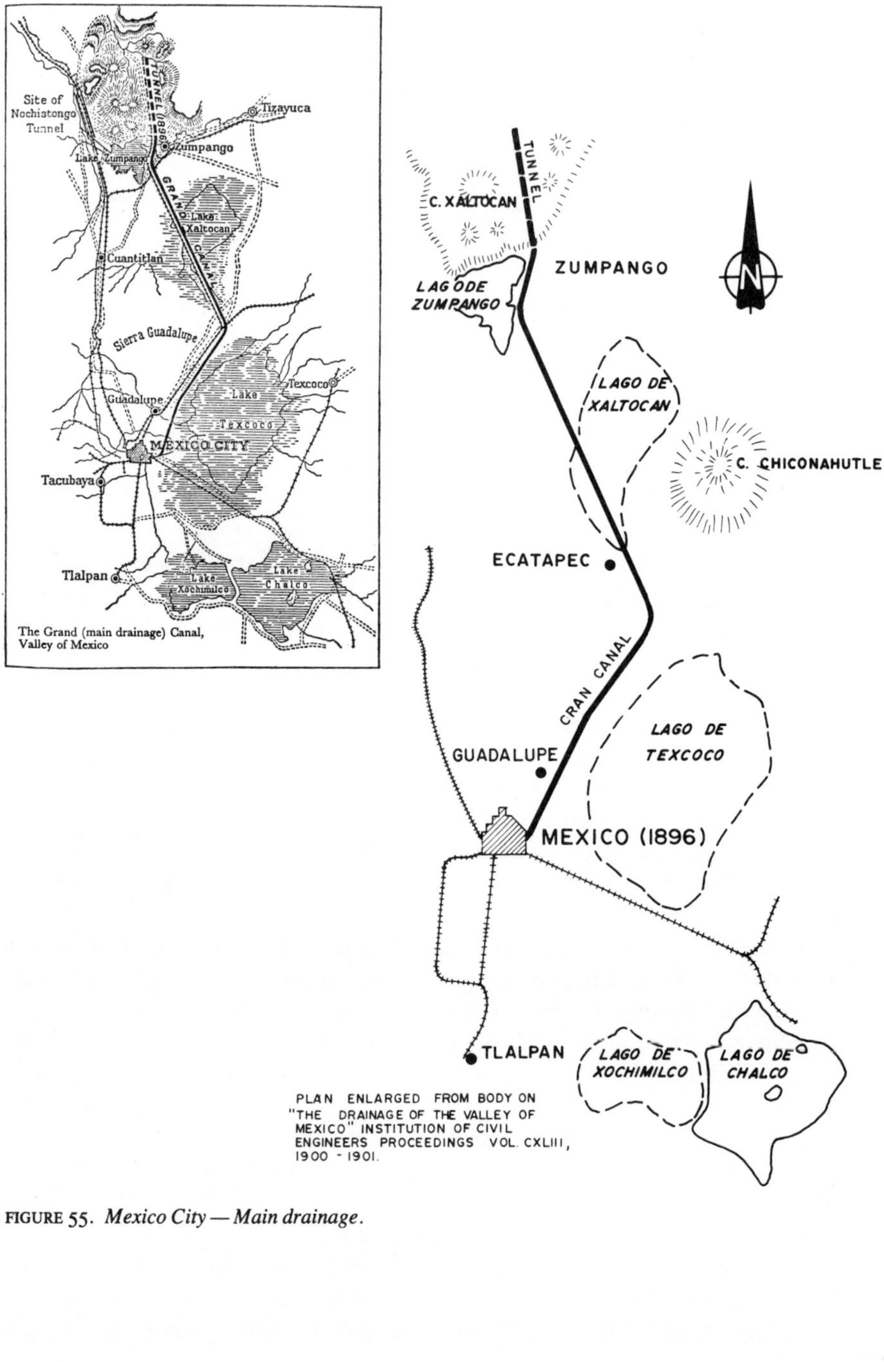

FIGURE 55. *Mexico City — Main drainage.*

on in Mexico City. **Figure 56**. This is due to the fact that the clay layers located between the ground surface in this city and a depth of about 300 feet below the surface are by far the most compressible clay strata which have so far been encountered in any part of the world. ... The subsidence due to pumping alone produced differential settlements by amounts up to 10 feet. The effects of these movements on public utilities such as the sewer system, is disastrous".

In my report (21.2.63) the importance of full site testing by actual work was emphasised. "Before any decision is made it is vital that shafts or large scale borings should be sunk along the line of the tunnel and the behaviour of the ground investigated in situ. Undisturbed samples of small size removed from the ground will not be the answer. ... the depth of the tunnel increases considerably in the latter stages and so may the pressure. It will be quite useless to take steps to investigate the section nearer Mexico City if even more stringent tests are not made in the deeper part of the tunnelling. Thus the dilemma is to be resolved as to whether this should prove a very easy matter of tunnelling or whether a monumental folly might not develop if at the last, portions of the work were found not to be possible".

One of the conclusions: "From the E.C.S.A. reports of their previous tunnels, Mexican engineers and workmen are fully competent to carry out such work, subject to advice on special precautions to be observed".

A sequel to this visit was the arrival at short notice of six Mexican engineers in London with the request to visit some compressed air workings and the manufacture of concrete segments. The leader of the party unexpectedly proved to be the Minister of Hydraulics who was also, most appropriately, the President of the Mexican Institution of Civil Engineers. These introductions led to the establishment by Kinnear & Moodie of a factory for making reinforced concrete segments in Mexico and the delivery by Markhams of Chesterfield of three tunnelling machines designed to work in the expected conditions. This ended my assignment. The subsequent story of the Mexico City Drainage and Metro must be sought elsewhere.

THOUGHTS ON LOCAL AUTHORITIES

Civil engineering contracts from their nature invite *the adversareal system of litigation* which involves the *Engineer* under the contract on one side and the *Contractor* on the other.

The single-handed independent consultant resembles a doctor, and is called in when needed. When all goes well he hears no more of his patient unless a relapse occurs. If his practice is wide he handles, often on a confidential basis, the troubles of contractors, consulting firms or officials, both British and foreign. At times he is appointed officially and addressed by Counsel as *Learned Arbitrator*! But in other cases, when both parties are willing, he can help to mediate privately in disputes

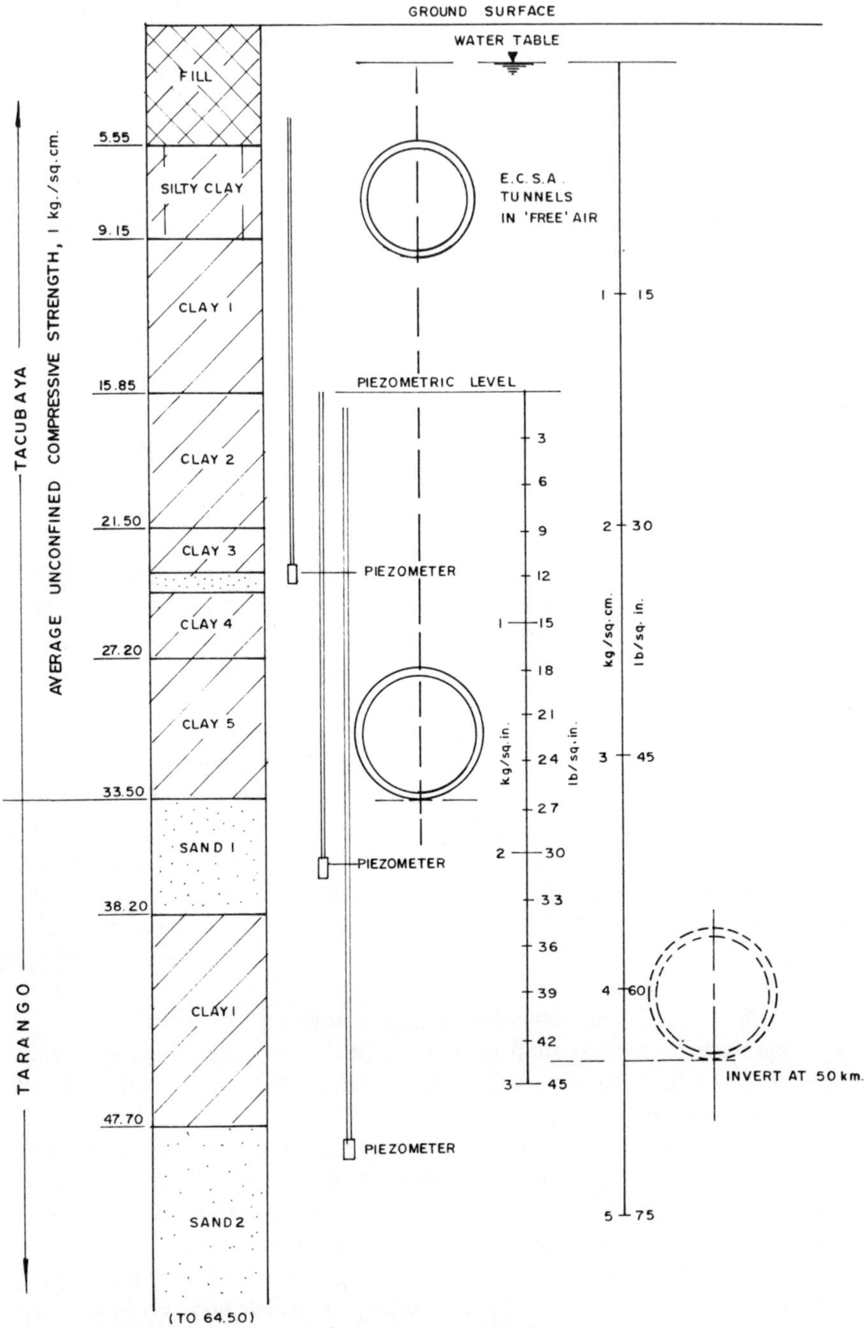

FIGURE 56. *Mexico City — Soil section at Tower Latino Americano.*

and so save accumulated legal fees. From all these experiences, the factor of human behaviour is often of more importance than the technical aspects. Money is the root of many quarrels which take the pleasure out of the engineering work.

This is more noticeable in sewerage works than in major tunnel contracts for several reasons. One is the fact that they are usually long though of relatively small diameter, and involve many shafts and manholes. They have not the financial muscle nor technical resources which are usually mobilised on the tunnels of first magnitude.

The engineer who works under local authority committees is responsible for spending and not for collecting revenue. He thus has little independence and less authority than the Town Clerk. Money for site investigation is grudged, as are payments for work done. As new tunnels are infrequent the borough engineer seldom has the continuous accumulation of experience which comes to the specialist contractor or consulting firms.

Contrasts in behaviour are met with between the engineer strong enough to use his own powers of decision under the contract and the engineer who would prefer to unload his responsibilities on to an arbitrator, who takes the responsibility (and protects him from surcharge by the District Auditor). This has little to do with tunnelling science but much to do with actual experience.

A city engineer of the first class (in two senses) wished to make a fair decision on a claim without wasting time and asked for advice. A long compressed air sewer in Bracklesham sands and Bagshot Beds was involved, leading to a claim including adverse physical conditions leading to excessive air consumption.

The sewer, 7 feet O.D. was driven by hooded shield and lined with bolted concrete segments. It was 6,000 feet long, but with only seven boreholes, some of them 70 feet off line and about 800 feet apart. **Figure 57**. The writer's report commented on the site investigation report: "The only criticism was that too much trouble has been taken over a great number of laboratory tests which have little interest to a tunneller. There is not sufficient emphasis on making clear the ground-water conditions. Permeability tests at only one end of the tunnel could be misleading".

The conclusion was that, up to a point, the experienced contractor should have expected the kind of actual physical conditions encountered. The amount of air in the estimate was far too low but the actual amount needed was more than should be reasonably expected, due to variations in the ground between the boreholes. The denseness of the sand from penetration tests does not reveal the permeability for the passage of compressed air, which depends upon the grading.

The City engineer paid for the cost of air consumption beyond the assessment by the writer, and left the contractor with the cost of the under-estimate below this assessment.

By contrast, a highly experienced contractor allowed for compressed air for part

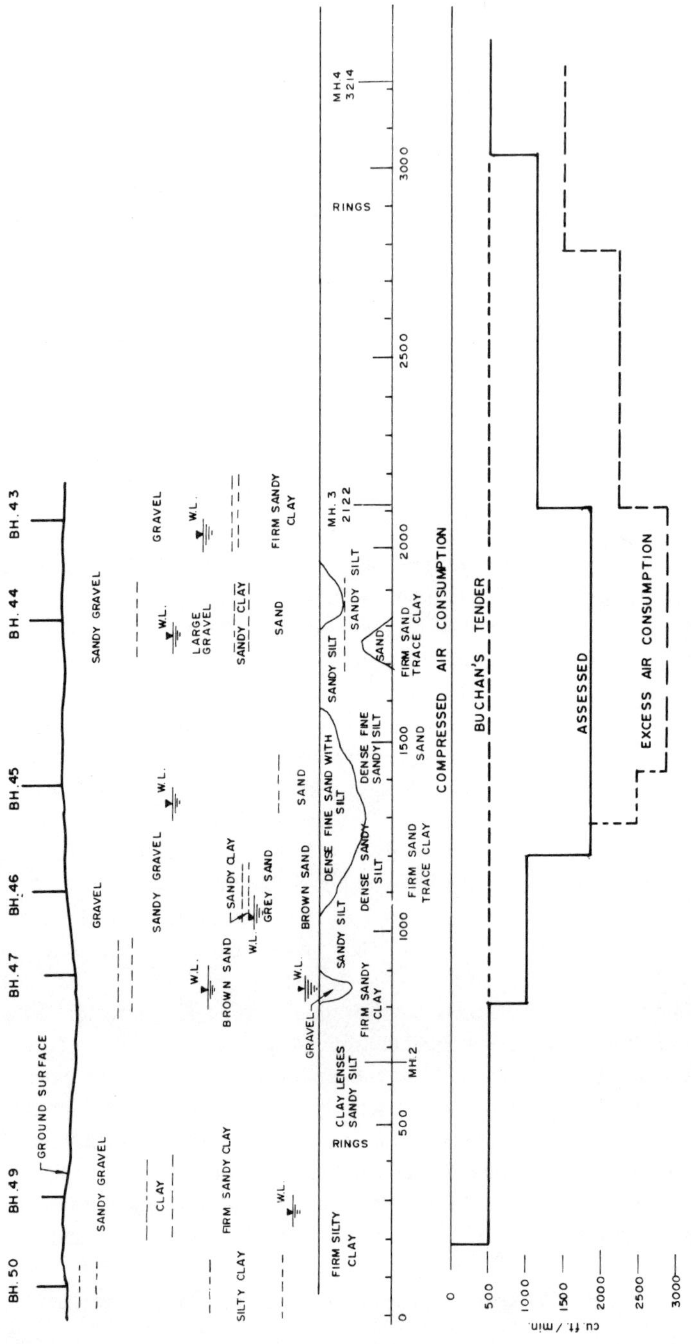

FIGURE 57. *Portsmouth sewer — Air consumption.*

of a long sewer tunnel. The Engineer persuaded him that it would not be necessary so the sum was deleted. In the end compressed air proved necessary for the whole length. This led to arbitration in which the contractor was successful, and the writer was a witness.

A second length adjoining this was subsequently let to a contractor who wished to diversify into compressed air tunnelling for the first time. Here once more tuition was provided and the work successfully completed and the contractor protected by the successful outcome of the previous case.

On the other hand a borough engineer was glad of protection from a formidable contractor who was bullying him by talking of the risk of miners being blown through the tunnel face. This may have occurred under the New York Rivers in shallow cover of soft silt, but is hard to imagine in a small sewer in sands and gravel under a main street.

Chapter 11

TUNNELLING IN CHALK

Chalk forms a natural transition from soft ground tunnels to rock. Among many definitions is one from the Code of Practice for Site Investigations: "a massive bedded white limestone of rough and *earthy* texture which may range from hard, clean fractured rock, through soft, easily abraded rock to white clayey or shaly marl".

The Cretaceous System (or Cenomanian) owes its name to the white chalk, widely spread across England from Flamborough Head in Yorkshire to Dorset, extending across the Channel into Northern France. The system contains many varieties other than chalk itself and also occurs in North America, Spain, Germany, Russia, and New Zealand as well as many other parts.

A wise tunnelling engineer should regard *chalk* with deep suspicion. It looks so safe when standing on 100 feet high cliffs or chalk pits, but it can be deceptive and can vary widely and suddenly in its condition. The British deposits have suffered folding, faulting and fissuring. They can be dry or heavily water-bearing and can soften and turn very quickly into mud when worked in wet conditions.

There are three main divisions in chalk which have been much discussed, worked upon, and tested during the *Hundred Years' War* to build a Channel Tunnel. The Upper Chalk can be of considerable thickness and contains abundant flints, sometimes of almost boulder size. These often occur in thick layers along bedding planes. The Middle Chalk has fewer flints. Both are apt to be heavily fissured so the Chalk is a valuable source of water. Many tunnels are driven in these, unlined, in very wet conditions for water supply.

The Lower Chalk on the other hand is devoid of flints and is more of a chalk-marl containing a small proportion of clay. In the pilot tunnels driven under the Channel in 1882 and in 1975 the Lower Chalk was found to be an ideal material for tunnelling by machine while fissures were almost non-existent. During the extensive surveys from 1958 to 1975 confirmation was obtained that the Straits of Dover were formed by erosion and not by faulting.

Chalk can appear solid but can easily deteriorate and become re-deposited as head, putty chalk, or Coombe deposits. It can be affected by ice-age conditions and reduced to chalky boulder-clay, often concealed by later deposits. The danger of abandoned unlined chalk workings has already been mentioned which, although appearing solid (provided that they are discovered) can flake away and allow running ground above to collapse into them.

The earlier railway tunnels in fissured chalk, driven from open portals could be made self-draining if the Engineer designed his grades for this purpose. Modern tunnels when driven from shafts often call for the exclusion of water and so involve a programme of grouting.

When grouting fissures it is important to ensure that the grout moves forward and does not flow back to soak the operator. This entails the use of expanding packers or other means to staunch the boreholes. In severe cases courage is needed to make the decision to stop progress and build a bulkhead against the face, through which injections may be made with more certainty.

THAMES CABLE TUNNEL (1967–1968)

This tunnel carries high tension cables at 400,000 volts under the River Thames from Tilbury to Gravesend. It is described in valuable detail by the Engineer for the work.(Haswell 1969). A tunnel as near the river bed as allowable would have been in Thames gravels and would have required a pressure of 40 lb per square inch. This was unattractive after recent research into cases of bone necrosis. A low level tunnel was decided upon in the Upper Chalk, lined with bolted concrete segments of 10 feet I.D. at 150 feet below mean tide level.

My advice was asked by an unsuccessful tenderer and some notes dated August 1966 survive. "Out of seventeen borings only Nos 7 and 6 penetrate below invert level and some barely reach axis level. It is necessary to extrapolate from Nos 6 and 7 and trust that they represent conditions at depth elsewhere.

"The report does not say whether the chalk is broken or fissured as the river bores were all bored by percussion. Land Bore No 2 has 80 per cent core recovery and No 13, on the south bank, has only 20 to 30 per cent the whole way". Then followed comments upon the possible water conditions, recorded in the borehole logs as Lugeons. One Lugeon is the equivalent of a permeability of the order of 1.5×10^{-5} cm per sec, or about the same as sandy silt. Penetrations, *N*. and Lugeons *L* varied from every borehole. The letter went on — "A shield is a necessity and whether this should escalate into a rotary digger is the estimator's guess ... If an optimistic progress rate is chosen it should be matched by money for pumping from the shaft".

Another suggestion was to leave a number of rings ungrouted and seal the length by a mortar garland at intervals. Then by grouting in a continuous operation, the water would be driven from one grout hole to the next as each is closed by a plug. This was done on this and on the next example.

The following advice was offered on the shafts where the design called for concrete bolted rings of 20 feet I.D. for the full depth, but surrounded outside by a circle of steel sheet piling shown as 28 feet 7 inch inside diameter. "On the north shaft the alluvium is nearly fifty feet thick with 15 to 21 feet of water-bearing Thames gravel below it, which will have an artesian head under the alluvium. On the south bank, bore No 13 has a bad core recovery down to the finish at −141.90 where the invert level is −138.00 ... I understand that the shafts are to be carried in sheet piling into the chalk and then by underpinning. It will be worth considering making the first ring of box of piles wider than need be so that a second, or, in the case of the south bank, a third section could be driven. This would give room for an intercepting drain round the lower concrete waling if used to catch the leakage through the clutches and so protect the lower work. Extra space would also give room for injecting with cement outside the lining before sinking in the higher level of the chalk".

In the Contract the Engineer had sensibly divided ground treatment and deep wells into separate sub-contracts and took other steps to avoid imponderables having to be priced so that such items would only have to be paid for as they arose. Thus no special advice was asked for or given on grouting or water lowering in this enquiry.

A year later I was asked by Mr Kyrle Clark, F.I.C.E., a Director of the old-established firm who had obtained the contract, to visit the North Shaft which was in trouble. **Figure 58** is from a sketch made at the time. Only one ring of sheet piling had been driven to 75 feet below ground level. The shaft, of 20 feet diameter bolted concrete rings was being sunk by underpinning inside the ring of piles. A blow had occurred under the foot of the piles which had barely touched the top of very soft chalk. Water was cascading down the shaft which was not yet caulked. As Mr Haswell wrote: "Unexpectedly the top 20 feet of chalk was found to be completely disintegrated, possibly due to permafrost". This was unexpected as in borehole 2 there was 80 per cent core recovery with no *N* values for penetration.

Mr Clark confirmed, in the discussion of the paper that work had been pressed forward "in the knowledge that delay would cost the employer around £20,000 per day" ... "Ground treatment would take time and owing to the somewhat lengthy contract procedure no sub-contractor had then been appointed".

Deep wells with submersible pumps had been sunk inside the shaft and met with artesian pressure, but not surprisingly this was not successful as well as being poor practice and hindering the work. Following discussions, as Mr Haswell records, "It was decided to instal additional and deeper wells outside the shaft". Five wells were inserted. "... The reduction of water pressure in the shaft was technically successful ..." They would be performing their proper function of drawing water away from the shaft and not into it. The discussion gave much useful comment from those who worked on the water lowering and comments from outsiders as to why freezing, monoliths or other methods had not been chosen.

Figure 58 also shows a sketch-suggestion for the South Shaft which was given to the unsuccessful tenderer, showing a three-stage method of piling, as in

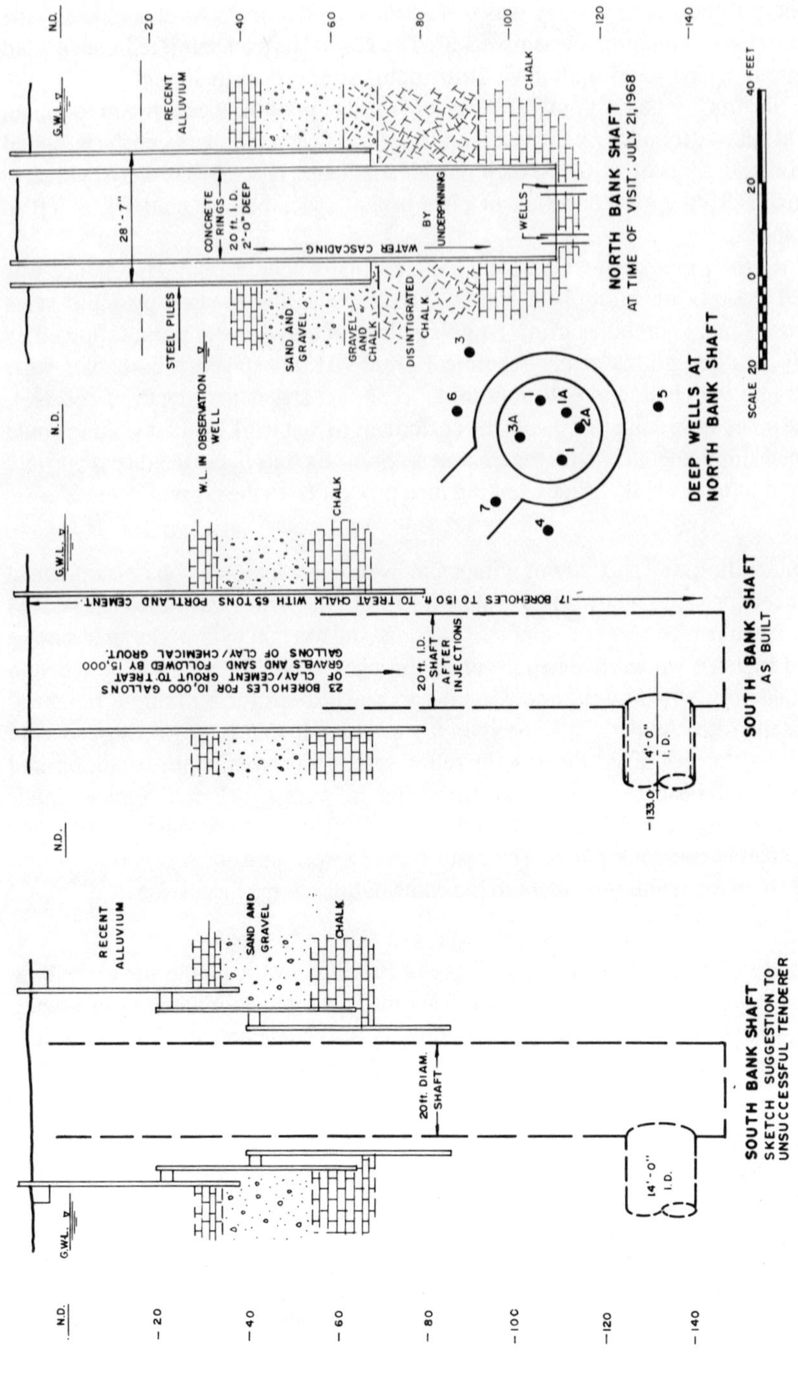

FIGURE 58. *Thames Cable Tunnel — Soils in shafts.*

borehole 13 the core recovery was bad down to below invert level. In the end the expectation of conditions was reversed. The North shaft encountered unexpected bad ground. The south shaft gave little trouble after proper treatment.

At the time of the visit only a single ring of steel piles had been driven for about 80 feet below ground level. This left only three feet clearance between the piles and the concrete rings of 21 feet 9 inch outside diameter. Discussions were held as to means of driving a second ring of piles in that space but the matter was left in abeyance.

When the Cementation Company had eventually been appointed, reliance was placed entirely on ground treatment. The gravels were treated by multi-stage injections in 23 boreholes using 10,000 gallons of clay/cement grout followed by nearly 15,000 gallons of clay/chemical grout. At this stage 17 boreholes were cased into the chalk and extended in 10 to 15 feet stages to a depth of 160 feet. These holes were sunk just inside the cofferdam so that work in shaft sinking could proceed simultaneously with the treatment. Some 65 tons of portland cement were injected into the chalk. "Shaft sinking then proceeded in the normal manner".

Tunnelling

The tunnelling was carried out with a Caldweld machine. After the machine had advanced for 1,700 feet with no ground injections the writer was called to a second visit. The machine was at a standstill after driving too far beyond a large fissure of increasing size which defeated grouting through boreholes. Sensible bold action was taken by withdrawing the machine to make room for building a bulkhead across the face. As the chalk stood up it was possible to take down rings so as to withdraw the machine. Pipes with relief valves were built into the brick and concrete bulkhead to control the flow. The injections included cement, sand, fly-ash, bentonite and bituminous emulsion. This delay took eight weeks. Two other such events took place. The rest of this successful work is described in the paper, with extra enlightenment in the contributions to the discussion.

BRIGHTON OUTFALL SEWER (1972–1977)

The outfall system for the main drainage of Brighton restricted the site to the base of high chalk cliffs near Portobello. This made the possible solution of towing welded lengths of steel pipe out to sea difficult to contemplate. The Consulting Engineers, Messrs Lewis and Duvivier, were faced with the problem as to whether a 7 feet diameter tunnel could be driven for one and a quarter miles out to sea in Upper Chalk at a depth below the limits of compressed air work.

Several borings at sea and a geophysical survey revealed reasonably hard chalk with no indications of pockets of weathered chalk or cliff falls reaching down to the zone of the tunnel. During the excavation of the low-level pump house, grouting tests were carried out from vertical boreholes after it had been found that horizontal fissures and bedding planes produced most of the water encountered.

In 1968 the problem was discussed with the Consulting Engineers following the lessons of Tilbury and other works and the undersea early Channel Tunnel pilot tunnels. The advice given was that the scheme was feasible and confirmed by letter. Some extracts follow: "the chalk should be pretty self-supporting but the volume of water would be considerable. The tunnel should be driven uphill with sufficient pumping capacity. ... I do not think that a shield is a necessity. It could help, but as the ground is too hard to push the cutting edge into it, the tunnel can be driven without a shield". (This shows a change of mind from Tilbury advice).

"I inspected some cable tunnels being driven for Charles Haswell and Partners with Kinnear Moodie as Contractors at Lewisham. In one case they had a great deal of water, but used ground-water lowering as the ground was impregnated with sulphuretted hydrogen and ammonia gases from decomposed dumps above, so that compressed air was not considered safe. In that case they did 6 feet forward in an 8 feet diameter tunnel in 24 hours using Kinnear Moodie bolted segments and no shield. In another area where they could use compressed air they did three 2 feet long rings in 12 hours, using compressed air tools for excavation.

"I understand that in your case you want a tunnel of 7 feet diameter giving 38 square feet in face area. This is rather constricted and only one man can work comfortably in the face".

The contract was let in 1972 to Kinnear Moodie Ltd., a firm of long experience and who had pioneered the Drum Digger and other innovations. They had just been taken over by a larger organisation with a different repertoire of experience and a different philosophy

Tenderers were given considerable freedom both in tunnel design and level, with all available information disclosed, on which to base their price. As at Tilbury the Engineer had attempted to reduce the imponderables. Probing was to be measured per linear mile of tunnel, drilling per mile of drilling, and grouting materials at net, plus a percentage to cover injecting, handling and waste. Pumping was left at the contractor's risk.

At Tilbury the choice of bolted segments was eminently suitable for a dry cable tunnel, but as they are costly to line to produce a smooth surface, the contractor in this case chose to use his own Rapid Lining for a 7 feet I.D. tunnel. This gave a smooth surface as it consisted of eight solid concrete blocks and a key. As the blocks have knuckled horizontal joints in order to give flexibility they need temporary steel erection plates bolted across adjoining blocks until the ring is keyed up. If grouting is delayed, plates must be left in position.

Unlike the Tilbury tunnel difficulties arose almost from the start. Large quantities of water issued from a number of fissures. Miners were in short supply and disgruntled both by the wet conditions and the need to drill and grout which spoiled their chance of profitable piecework. The small size of the tunnel and the obstruction of the body of the shield gave little room for mining and drilling at the same time, while the grouting of rings at a distance behind the face interrupted the service to and from the face.

It soon became evident that the work was badly underpriced, especially as the items for drilling had been marked as *included*. Possibily this was the work of a quantity surveyor and not the agent likely to do the work. To add to their troubles the new parent firm suffered from a financial crisis on work abroad which starved the contract of much needed funds.

By a coincidence the equally experienced and reputable firm which had driven the Tilbury tunnel so successfully had themselves been taken over by a large firm with no engineering background. This led to the shedding of highly experienced directors. One of these, Mr Kyrle Clark, well known to me, had become an independent consultant and so was recommended to the Consulting Engineers to reinforce their supervision in such difficult cases. This led to discussions and three visits with him to see what needed doing.

The record of all this activity covers many pages and can only be touched on in this account. The attempted rhythm between mining and drilling was upset by differing rates of progress in each type of work and length of drilling, with time for injection.

Various patterns of drilling and methods of injection were discussed including the pros and cons of using specialist grouting and drilling sub-contractors. Work proceeded slowly with difficulties in sealing packers into fissures; some were filled with putty chalk when first met but water washed this out and fresh treatment was needed. My letter dated 16 July 1973: "There appears to be solid chalk relatively speaking to the left of the face but considerable fissures to the right. ... My notes say that twelve shifts were spent grouting from two holes 30 feet into the fissures and 88 tons injected in varying quantities in one and 34 tons in the other". This is one example.

Owing to financial stringency there were defects in pumping capacity and other aspects of plant and equipment, noted in the visit on 24 July. Other comments were made after 14 August with the intention of helping the local director with his head office superiors, which may convey a message. "As this (later) fissure connects to the sea bed, according to your divers, you might be held remiss if you did not take all precautions to ensure that the tunnel passes the area safely. Time spent now is as nothing as compared to trouble if care is not taken ... You should insist that two widths of floor boards should be brought up to the face and secured. If you have trouble and there is no safe and rapid retreat for men the Factory Inspector could react violently. In any case the floor boards are missing for most of the way. This is thoroughly bad practice and should be dealt with.

"In such a case as this where difficulties occur and need to be overcome by changes and experiment, especially in ordering plant, material and outside assistance, I have found from experience that this places too great a load on an agent on such a small scale job, a long way from his base. For when he should be down the tunnel he is on the telephone seeking help, and vice versa. If this was my job as a working Director I should sublet most of my other work to a colleague and give considerable undivided attention to the site in order to support the agent and foreman by helping to provide not only advice but influence in order to extract all support needed from inside or outside the firm. This is better than back-seat driving from an office".

This may have little to do with technology but the human factor at all levels cannot be ignored. Sometimes, under contract conditions, reluctance to press on may have strategically motivated reasons. However, soon after these visits the tunnel came to an abrupt halt when the main Contractor went into liquidation. A long pause ensued.

Eventually Kinnear Moodie were again taken over and revived as Kinnear Moodie (1973) Ltd. Through the enlightened behaviour of both Engineer and Client a new contract was negotiated. In this second contract grouting was paid for under four headings. Drilling paid in linear metres, setting up of plant, etc., per number, grouting material as before, and the operation of grout pumps per hour. Pumping remained at the contractor's risk. With this new injection of financial strength the work went ahead and overcame more difficulties than those met at Tilbury.

Nine diffusers had been bored into the sea bed from a jacked-up platform along the last 300 metres of the 1,830 metre tunnel and the final work was to drive out to connect to them. My last visit was at half-way when the chalk was in good condition which raised doubts as to the value of the shield. Our previous various advices had been accepted and acted upon so the final account can be left to Mr Kyrle Clark: —

"Generally, after the restart with the *new* contractor, the system was to drill and grout four 40 feet holes on the 8 hour night shift. This worked reasonably well until the ground deteriorated for the last 800 feet. Then the friable chalk necessitated grouting through bulkheads. Lengths drilled and grouted from each bulkhead were about 60 feet but only about 30 feet progress per bulkhead was obtained. Each bulkhead took about a week cycle. The maximum take of grout in a bulkhead was 80 tons.

" The *rapid* rings were probably economic overall, though a bolted ring would have made construction easier. A change to cast iron rings was made for the last few hundred feet when the *rapids* appeared too heavily loaded and were going out of shape.

" A shield was used throughout. My feelings were anti-shield at the start. I know not how desirable it was. Everyone else is convinced it was necessary. The foreman and gang said that they would not proceed without one, which clinched the argument on this, to miners, not a very attractive job. The same sort of reason resulted in the short diffuser connections, cleaning out, etc., taking a long time".

The tunnel was thus completed in 1977 successfully, to the natural relief of those who had always maintained that it was feasible.

A different case arose some years earlier when widening the entrance to the railway tunnel at the entrance to Lewes Station behind the brick head wall. The original work had been in dry hard chalk. When a timbered heading was driven alongside and outside the brick lining the hard chalk had turned in the course of time, from exposure to water and fumes, to a soft putty.

Chapter 12

TUNNELLING IN ROCK

Directly after the war, Soil Mechanics Ltd. carried out site investigations for the Scottish Hydro-Electric Programme, followed by rock tunnel contracts for Mowlem (Scotland) Ltd. This brought the first personal confrontation with rock since geology courses at the Imperial College twenty-five years before.

Young engineers spend day after day prowling up and down their particular tunnel contract, absorbing every small detail. When they rise to a high executive standing, directing a number of simultaneous and quite different contracts, their visits to sites are of necessity more fleeting. To overcome this handicap I followed the advice of Sir Henry Japp: "Hire someone who knows how", by taking as my personal assistant for rock tunnels one of our younger engineers. He had returned from completing a difficult Hydro tunnel at Machkund in South India (Gould & Newport 1954). During the war he had spent several years on the Burma front as a sapper, so explosives were no novelty to him.

On a holiday in Sweden in 1947 I made a long-standing friend in Sven Platzer, who was founding his firm of Widmark & Platzer after years with Sentab. He gave me good advice on rock, I gave him advice on *soft* to our mutual advantage. Then, after the war, international conferences multiplied and firms even began to pay for some of their staff to travel abroad and exchange experiences. The 1948 International Conference on Large Dams in Sweden gave all who went an insight into their unusually solid uniform rock. After years of stumbling over the rough floor boards in circular soft ground tunnels it came as a surprise to be driven to the face of an unlined sixty foot wide, thirty foot high hydro tunnel — in a motor coach!

TUNNELS IN THE CARBONIFEROUS

In the early 1950's I was responsible for a number of tunnels in the Carboniferous, mostly in the South East Scottish Coalfields. These were to make access adits so that the coal miners could be left free to carry out their function of hewing coal. The ground varied from shales, sandstones or blaes – defined in Scotts Penguin Dictionary of Civil Engineering as "(a Scots term meaning Blue) Black carbonaceous shaly material in the Coal Measures, some of which are used for brick making". The tunnelling took the simple form of drilling and blasting with steel arch rib supports, using Eimco loaders or Joy Sullivan slushers for loading.

A detailed study was made of twenty of these tunnels which showed many differences in man-hours for various functions, output and progress although carried out by one organisation, often with identical gangs in geological conditions which might appear identical to a layman. This is a reminder of the wide choice open to estimators and of the effect of the unexpected.

The Whitehaven dipping tunnels need more explanation.

In 1954 the new firm of Solway Chemicals, near Whitehaven needed supplies of anhydrite for their processes. To their surprise they had unknowingly built their plant above a considerable deposit. So they started to drive sloping access tunnels at 1 in 7 by direct labour, using local miners.

The start was in Carboniferous shale, 13 feet wide, 9 feet high lined with arch ribs. After about 500 feet in the first tunnel the ground turned to soft mud with inflows of water and sand, which defeated them. We then took on a contract to complete the twin tunnels, using the local miners. The first step was to train them in soft ground methods and to line with bolted concrete segments until the bad length was passed, and then revert to ribs. This came as a surprise to the miners, 95 per cent of whom had no knowledge of segmental rings and thought that ribs were the only way to line a tunnel.

Both drives were completed for 2,700 feet to the anhydrite and then driven another 300 feet to explore the deposit. The work took one year. Solway Chemicals had provided conveyors in the first tunnel but two stages of winches were used in the second for winding. The conveyors were useful but cumbersome. They delivered spoil faster when working but caused more unproductive time.

During this time tenders for Scottish Hydro tunnels were unsuccessful as, at the mention of the word *overbreak*, elderly directors clapped on considerable sums, regardless of how much had already been allowed.

DEEP SHAFT SINKING (*an interlude*)

After the coal mines had been nationalised, fresh deep shafts were needed. The National Coal Board were quoted as saying that only German firms could carry out this work, with the exception of Cementation Co., as no British firm was competent to do so. This we could not accept. In January 1954 I visited the works of Mowlem Construction Co. in Uganda and Kenya and took the chance to fly on to Johannesburg.

With the help of introductions from Holman's Ltd. I was able to talk to the Chief Engineers of six gold mining firms. Unfortunately the first big shaft sinking programme in the new Orange Free State field had been completed but I was able to descend, rapidly, the newly sunk, mile-deep shaft at Welkom. Each expert had his own ideas. Some liked square shafts, some circular. Some liked big grabs under the galloway stage, others small ones. It seemed not to matter which was adopted as long as it was sensible. The collection of much data led to success in obtaining contracts to sink three deep shafts near Dalkeith. But they were promptly cancelled for lack of planning permission.

ROOF BOLTING AT OBAN CABLE TERMINAL

1954 marks the construction of an underground apparatus tunnel as the terminal of the first trans-Atlantic Telephone in the slope of a hill south of Oban. The Engineer had investigated the site but, as so often seems to happen, after the contract was let the position of the chamber was moved a few hundred feet away from the area tested. So Soil Mechanics Ltd. bored diamond drill holes at each end of the future tunnel.

Doctors Golder and Meigh then produced an internal report on "The Theory and Practice of Roof Bolting with Particular Reference to the G.P.O. Tunnel, Oban". At that date, in 1954 it is interesting to read: "Since 1947 roof bolting has been fostered by the U.S. Bureau of Mines and is now used extensively both as an adjunct to timbering and more frequently as a complete method of roof suspension. ... Roof bolting in tunnels is now being adopted in American Civil Engineering work. ... The recent use of roof bolting in British collieries has been described. ... No reference to its use in British Civil Engineering practice has been found". So the work at Oban was somewhat experimental. **Figure 59**.

After several pages on practice and theory the Report dealt with the Oban tunnel. It said that the main tunnel should be entirely in phyllites (which was to prove a new personal experience). There are several definitions of this metamorphic rock "produced by the continued action of stresses on slates" with the possibility of strain-slip cleavage. Core recovery was nearly 100 per cent in the lower part of the 30 feet thick conglomerate from the surface but the phyllites were more variable. In the siltstones and gritstones the core recovery varied from 40 to 60 per cent. The report recommended the use of rock bolts 8 feet long in the main chamber and 5 feet 6 inches in the adits and this was adopted.

A provision had been made in the contract for four very heavy ribs (from memory, 18 inches × 6 inches or heavier) welded in lengths to form a polygonal arch. The pilot tunnel revealed much worse conditions than had been expected from the original adjacent site investigation. Then occurred an incident in contrast to the conflict at Derbendi Khan between the parties over extra ribs, described later. It also shows the value of mutual personal contacts over the years. On telephoning to the Consulting Engineer with the advice and request for him to authorise steel arches at every 4 feet, the instant reply was: "If you say so, that is good enough for me".

The concrete lining was placed by pumping, using a complete travelling steel shutter which needed stiffening after the first pour. This early case of roof bolting taught the practical difficulties of drilling and installing bolts quickly enough after a blast. The pilot heading improved the chances of quick support. There is a

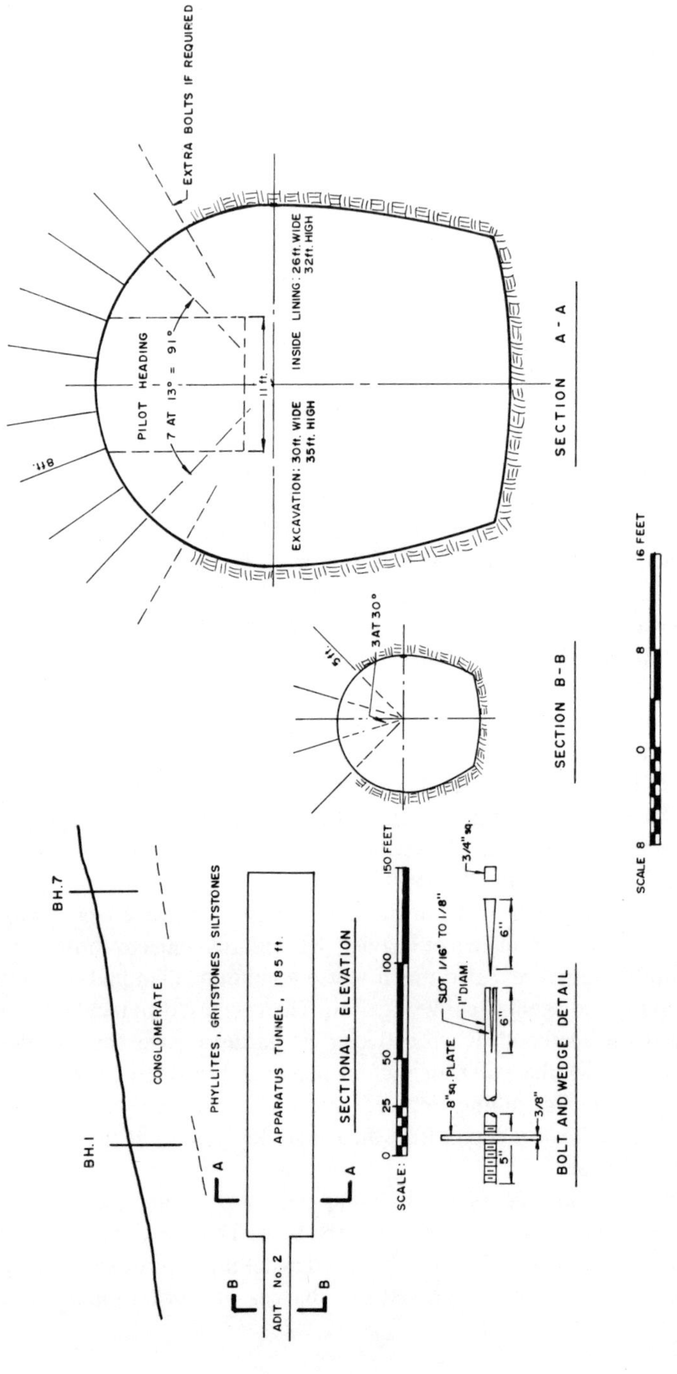

FIGURE 59. *G.P.O. Tunnel, Oban—Roof bolting.*

temptation to choose too short a length of bolt in this successful but slightly illogical method of pulling oneself up by one's boot strings.

COO UNDERGROUND POWER STATION, BELGIUM

The Oban exercise proved its value when, in 1970 there came a request from a British Insurance Broker to assess and evaluate a Belgian claim. This concerned the collapse of part of the roof of an underground Power Station or *Centrale* at Coo in Belgium. The massive documentation was in French, since returned to the client.

This Centrale was being driven, as at Oban, in phyllites and quartzites with quartzo-phyllites. It was 128.20 metres long with a roof arch of 12 metres radius. The entry was by one adit at crown level at the Pignon Nord or North Gable End, with another adit at one side in the centre of the length.

The first phase consisted of driving a gallery for the whole length in a somewhat random fashion varying from 12–14 metres wide. **Figure 60** shows the method adopted for support. Heavy timber props were placed at intervals and roof bolts 2 metres long held up steel mesh for spraying with concrete.

The intention for the next phase was to work down to the floor level by about a metre at a time, extending the roof supports until the retombée or springing was reached. Then the roof collapsed for a length of 10 metres starting from 6.50 metres from the North Gable. The overbreak reached 3 metres with a fall of 60 cubic metres. Fractures were reported as well as opening up of the thin-leaved beds of shale, gneiss and quartzite on either side of the collapse and water issued from a water-bearing fissure.

The Mines Department appeared to have authority to approve or instruct on steps to be taken. On their insistence the roof for 51.5 metres from the North Gable was supported by reinforced concrete arches at close intervals with reinforced concrete ribs running along the tunnel. A much more intensive pattern of roof bolting was insisted upon for the whole length of the tunnel. Patterns of 27 bolts occur at every 1.40 metres and these vary in type and length from 2.50 metres up to no less than 10 metres. The permanent arch had to be carried down to springing level before the floor level was lowered.

A lengthy site investigation report by the University of Liège was included in the main dossier. It seemed as if the line of the chamber had been diverted from the original intention as a line of boreholes starting from the west side of the South Gable passed outside the east wall of the chamber at 30 metres south of the North Gable and so missed the area of the collapse. **Figure 61**.

The report stated that, "the quartzite beds were never more than 0.50 metres thick and the phyllites 1 metre thick, and that these confer on the massif an altogether homogeneous condition and in consequence show a satisfactory uniformity from the point of view of the mechanical properties. But at the North end

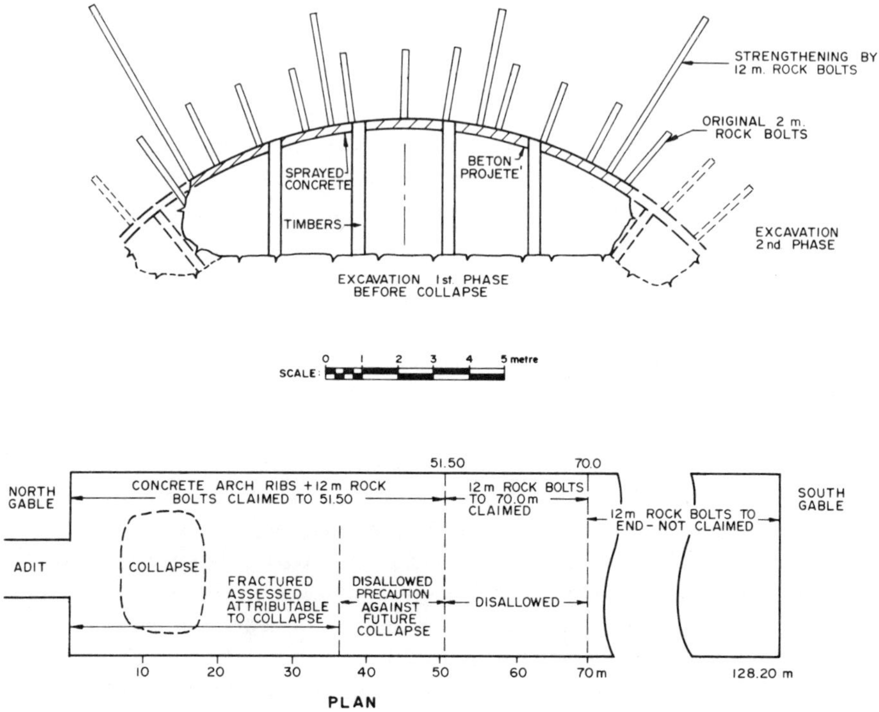

FIGURE 60. *Coo Underground Power Station, Belgium — Roof bolting.*

should be added the risk of slipping, following the bedding planes inclined to the South".

Many sheets in the dossier disclosed details of the tender items, several versions of variations and the final certificate details. Those disclosed the fact that the Contractor had been paid the whole of the expenditure subject to a credit on the insurance claim.

My first assessment was that only 30 metres of the 51.5 metre length of the concrete work could be attributed to the disturbance of the rock. I then asked the Ingenieur des Mines as an independent party whether he thought this a fair assessment from his intimate knowledge of the work. In a lengthy dissertation he finally concluded that 36.5 metres should be allowed for the measures to restore the situation.

My assessment covered several sheets of details of many items, some awarded 100 per cent, some nil; and the main concrete item 70 per cent. The claim beyond 36.5 metres I disallowed on the logic that a "party should not be penalised for damage which he has not yet been given the opportunity of causing".

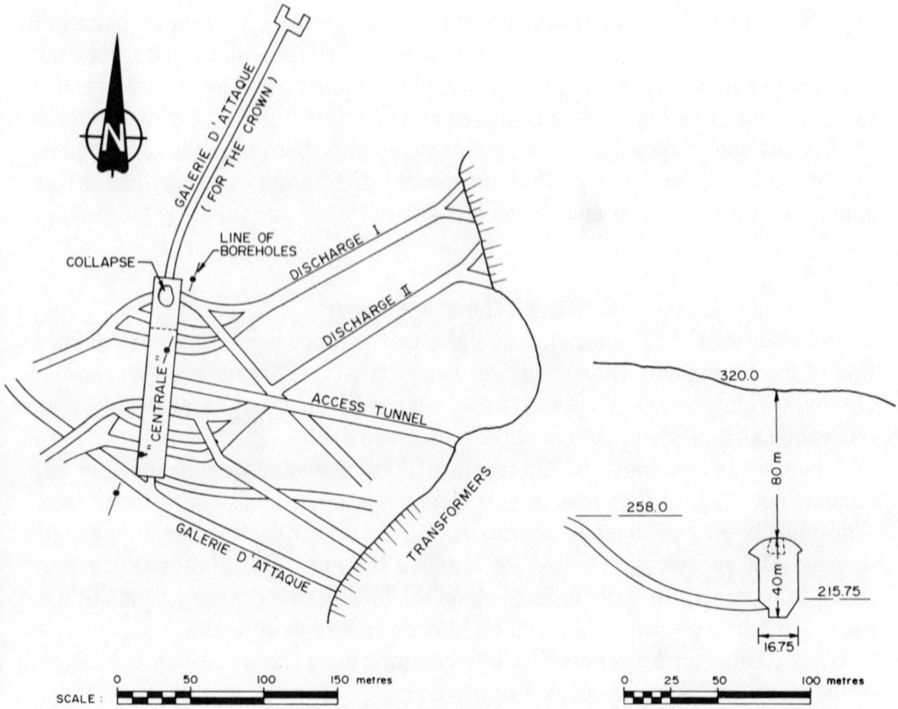

FIGURE 61. *Coo Underground Power Station — Site Plan.*

KARIBA NORTH

Two years after studying the events at Coo Power Station came a brief involvement in an impending crisis during the construction of the Underground Power Station on the north bank of the Zambesi River at Kariba. This involvement consisted of listening to the account of the Contractor's mounting difficulties and a brief, unsuccessful attempt to help towards an amicable solution.

Some account of the affair was published in the New Civil Engineer, 8 February 1973, including comments upon the Report of the three-man panel of experts. This was headed by Brigadier E.J. Edney, F.I.C.E., who had also practised as a Barrister-at-Law, Dr Charles Jaeger, M.I.C.E., the Swiss authority on underground power stations, who was appointed by the Contractor, and Dr John Knill, M.I.C.E., Reader in Rock Mechanics at the Imperial College, appointed by the Consultants.

As at Coo, after a successful start, rockfalls occurrred both in bolted and unbolted areas of rock. Once again, as in previous examples, the site of the Chamber had been moved some distance into the hill from the original borings and exploratory adits. The panel found that the borings were not factually accurate and conditions could not reasonably have been foreseen. Before contractual arguments had reached anywhere the Contractor went into liquidation. The work was eventually assigned to a Jugoslavian Contractor.

This example emphasises the way in which human behaviour, as suggested in the Introduction, and the general ambience of differing contracts can affect solutions, both technical and financial, with as many variations as the ground itself.

THE SHIELD IN SOFT ROCK

Useful information and some wise thoughts can be found in the paper, "Construction of the Monmouth Tunnels in Soft Rock" (Perry & Thornton 1970) and the Discussion (July 1971). Twin road tunnels were driven through a projecting spur on the outskirts of Monmouth to improve the A40 road.

A note on the geology and the results of five boreholes were included in the contract documents. They revealed clay and loam, marl, sandstone and marl, and sandstone dipping at about 40 degrees across the line of the 185 metre long twin tunnels. The specification called for full-face driving, with reinforced concrete lining to be never less than 30 feet behind the face in case of deterioration of the marl. The tenderers had to submit their intended methods of work.

When starting on the portals the contractor took great alarm at the broken nature of the steep hill-face and decided to use a shield. This was partly inspired by a series of collapses of very broken rock under a built-up hill at Newport, after which a shield had been introduced. The contractor (Cleveland Bridge & Engineering Ltd.) had diversified into soft ground tunnelling for many years and their extensive workshops soon produced a strong and ingenious arched shield. **Figure 62**. The consultants agreed to its use but at this stage declined the invitation to pay for it as the method was contrary to the tender submission.

My contribution to the discussion of the paper can save fresh words by introducing two other cases: –

"The comments in the Paper on the use of a shield in rock tunnelling are worth pondering. In effect, it is a combination of a moving rib and a drilling *jumbo* and can be a protection against the unexpected, but only up to a point. I saw the shield in action on two occasions because I was asked by the consultants to give an opinion on the interpretation of the bores when the discussions first arose as to what the ground was going to be like. As it says in the Paper, the shield was not included in the tender offer but was decided upon when the loose ground at the portals caused misgivings.

"That has an interesting psychological and financial implication. But as the question arose, one side saying that the ground might be very bad and the other hoping that it would not, I suggested that a coloured photograph should be taken of

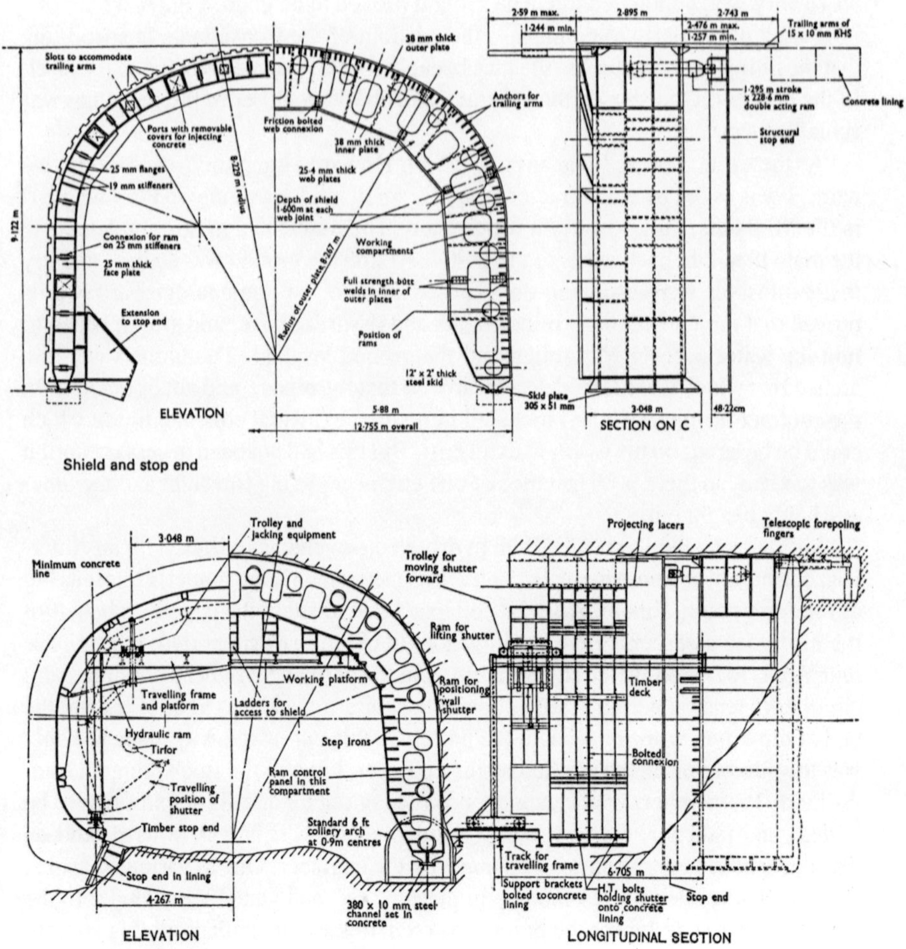

FIGURE 62. *Monmouth Road Tunnels — Roof shields. (I.C.E.)*

every length because such photographs show up the different coloured strata and would show finally where the truth lay by scientific proof. I saw these later and it was a very well conducted job. The ground proved to be good. **Figure 63**.

"I wish particularly to emphasise the wisdom of the Consultants in specifying that the lining must be close to the face because of the possible swelling of the marl. In the end, with the shield, the contractor lined much closer to the face than was actually specified.

"At the same time as I was involved with Howard Humphreys & Sons on this work, I was asked by a Contractor to visit a small but long tunnel in Permian marl in the Midlands, built to carry a water main. The tunnel had to be completed and the main then laid backwards to the portal. At first the ground was so hard and dry that explosives were used and dust caused trouble, but the geologist afterwards proved that there were many minor faults and fissures above, and after it began to rain the water percolated through and the ground swelled. The tunnels were on arched ribs which were forced downwards so that the pipes could not be entered. In the contract the engineer had included items for segmental concrete lining which could be ordered, on his word, at extra cost. But this had not been necessary until it was too late, so there is an instance of one engineer having foresight and the other one being caught out.

"Similarly on the Derbendi Khan hydro-electric scheme in Iraq, with an American consultant and an American contractor, there was a large tunnel, almost as big as the Monmouth Tunnel. After its collapse I had to study the papers on behalf of the insurance assessors. Here again was a tunnel in marl supported on ribs. The tunnel was to be completed in six months and the permanent light steel lining, kept circular by tension wires, acted as concrete shuttering and had to be laid from the end of the drive, working back to the portal. In this out-of-the-way place not only was the number of ribs controlled by the engineer, but also the stock piling in Iraq. As far as I remember, in correspondence the contractor asked for extra ribs to be issued, and paid for, for use at 0.75 metres centres. This was refused, and an insistence at 1.5 metre centres was made by the engineer. Due to various reasons, the tunnel was open for ten months in place of six, and water percolated into the marl once more: a rib near the portal was overstressed and collapsed; this brought down the next and the next etc, in a sort of domino action.

"This raises the question of who is to take the risk and who is to pay. For instance if a shield had been included in the Monmouth tender, would the tender have been accepted?. There it was a sensible precaution, whether or not it might have been possible to complete the tunnel, as specified, on ribs without a shield and that is a point which the authors might like to ponder".

Tunnelling progress was remarkably rapid, but as the actual mining work had been sub-let the sub-contractor profited handsomely while the main contractor was left with the expense of two shields and the heavy shutters to make it possible to jack the shields from green reinforced concrete lining.

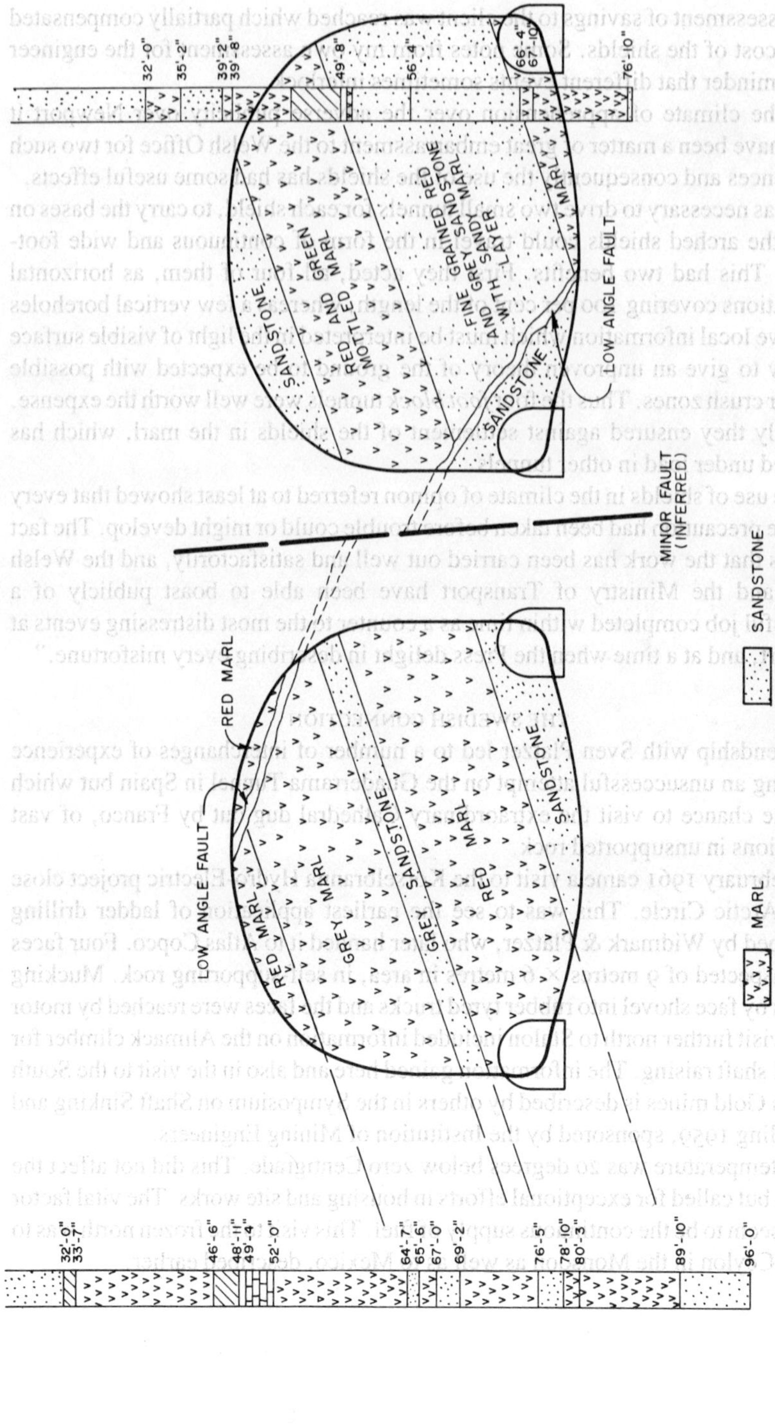

FIGURE 63. *Monmouth Road Tunnels — Cross section of faces.*

An assessment of savings to the client was reached which partially compensated for the cost of the shields. Some notes from my own assessment for the engineer are a reminder that different events sometimes interlock.

"In the climate of apprehension over the adverse publicity over Newport it would have been a matter of great embarrassment to the Welsh Office for two such occurrences and consequently the use of the shields has had some useful effects.

"It was necessary to drive two small tunnels for each shield, to carry the bases on which the arched shields could travel in the form of continuous and wide foot-blocks. This had two benefits. First they acted, all four of them, as horizontal explorations covering 100 per cent of the length, whereas a few vertical boreholes only give local information which must be interpreted in the light of visible surface geology to give an unproven theory of the ground to be expected with possible faults or crush zones. Thus the four *foot block* tunnels were well worth the expense. Secondly they ensured against settlement of the shields in the marl, which has occurred under load in other tunnels.

"The use of shields in the climate of opinon referred to at least showed that every possible precaution had been taken before trouble could or might develop. The fact remains that the work has been carried out well and satisfactorily, and the Welsh Office and the Ministry of Transport have been able to boast publicly of a successful job completed within time as a counter to the most distressing events at Newport, and at a time when the Press delight in describing every misfortune."

THE SWEDISH CONNECTION

The friendship with Sven Platzer led to a number of interchanges of experience including an unsuccessful attempt on the Guaderrama Tunnel in Spain but which gave the chance to visit the extraordinary Cathedral dug out by Franco, of vast dimensions in unsupported rock.

In February 1961 came a visit to the Korselbranna Hydro-Electric project close to the Arctic Circle. This was to see the earliest application of ladder drilling developed by Widmark & Platzer, who later handed it to Atlas Copco. Four faces were inspected of 9 metres × 6 metres in area, in self-supporting rock. Mucking out was by face shovel into rubber tyred trucks and the faces were reached by motor car. A visit further north to Stalon included information on the Alimack climber for vertical shaft raising. The information gained here and also in the visit to the South African Gold mines is described by others in the Symposium on Shaft Sinking and Tunnelling 1959, sponsored by the Institution of Mining Engineers.

The temperature was 20 degrees below zero Centigrade. This did not affect the tunnels but called for exceptional efforts in housing and site works. The vital factor would seem to be the continuous supply of fuel. This visit to the frozen north was to lead to Ceylon in the Monsoon as well as to Mexico, described earlier.

THE CEYLON HYDRO ELECTRIC SCHEME

In 1959 the contract for Stage B of this scheme was obtained by a joint venture of Skanska Cementgjuteriet with Widmark & Platzer, under the title of Skanska-WP. A horizontal tunnel, 20,000 feet long, was to be driven in gneiss and schists, described in one report as of quartzite bands in Charnockite. The Ceylon Highlands are of the same age as the Scottish Highlands but the slopes are covered with tea, rather than heather.

The tunnel led from the recently built Castlereagh Dam to discharge through a steel tube (by Japanese Contractors) to a Power Station in the open on the edge of the Norton Dam, 500 feet below the intake.

The contract followed the old Hudson style of 50 years ago, which placed every conceivable and inconceivable risk on to the shoulders of the contractor, in contrast to the modern I.C.E. Conditions.

The Government of Ceylon had taken over the traditional powers of the Engineer (in this case Coode & Partners): "The Schedule Rates shall be deemed to cover and provide for all such risks unless the Government and Consulting Engineers shall deem otherwise".

In 1961 I was asked by Platzer to introduce them to an English Solicitor: my choice was good as he later became President of the Law Society. In due course our efforts led the Government to "deem otherwise" over the risks which did arise. Some of these extended the length of the contract beyond tunnelling itself. For instance, after independence the port of Colombo continued in a condition of strike or go-slow so that ships had to lie off at sea for three months before unloading. The ship carrying the crushing plant gave up, turned around and unloaded it at Djibouti! The build up and start of the work took time.

I was sent the documents and Contract Conditions to study and comment upon and during 1962 and 1963 different problems were discussed by letter and in visits to London. Then in September 1963 I visited the site when a critical juncture had been reached. All the tunnelling had been completed except the length south from Adit O which had been in constant trouble.

As local talent was almost non-existent, Widmark & Platzer had been allowed to import Filipino miners who had been highly trained by them on two hydro-electric contracts in the Philippines. The progress in all faces had been on average, 100 feet per week but when the southward drive from Adit O had covered 650 feet by July 1962 highly adverse physical conditions were encountered. The story from July 1962 until September 1963 became very involved.

From chainage 1,675 for 180 feet the face had changed suddenly from solid rock to three areas: solid rock on one side, broken shattered rock and areas of very soft clay made from decomposed rock above and down to invert at times, with plentiful water in addition. There seemed no reason to anticipate the decomposed rock at

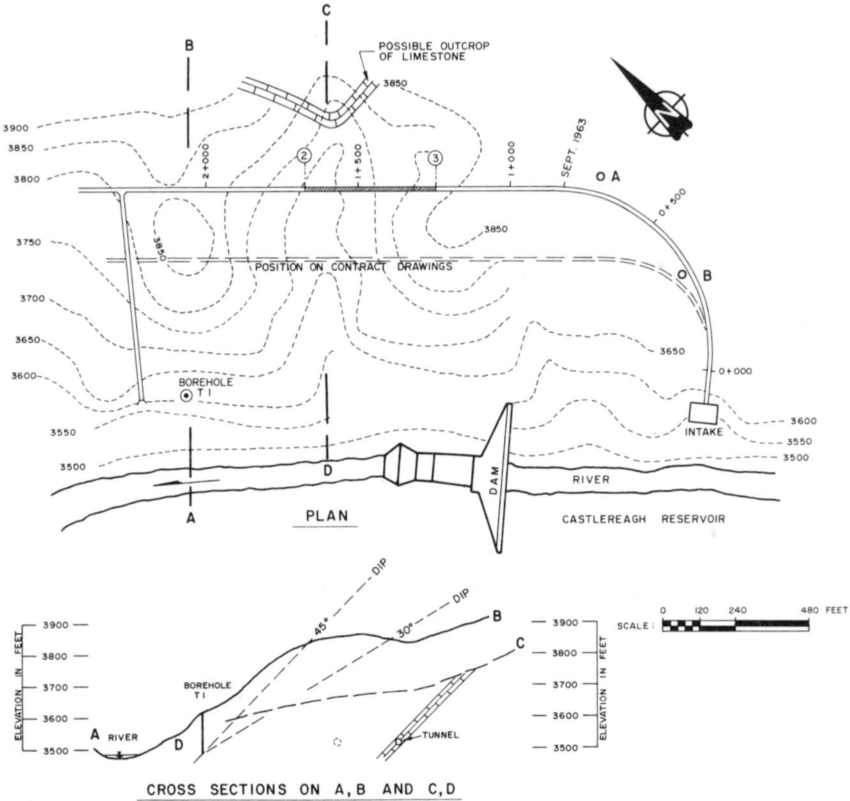

FIGURE 64. *Ceylon Hydro-Electric Scheme — Plan and section at end of tunnel.*

such depth from the tender information. The bill of quantities for the whole 20,000 feet only allowed for an item of 30 tons of ribbing and corrugated sheeting.

Progress became very slow. Benk bars pitched as spiles could not be fully driven home, while some had to be tailed down from the ribs behind. Although they would protrude into the future concrete lining it was considered too risky to disturb them, or the rock and timber packings which had been stuffed behind them.

Long pauses ensued, interspersed with short drives in better rock followed by collapses of the face in bad conditions which led to softening of the floor areas supporting the ribs. Arguments and recriminations passed slowly between London, Malmo and Colombo without the Principals being visible on the site. Eventually after many refusals, permission was given to concrete the sub-invert and eventually to allow the internal diameter to be slightly reduced in these lengths to give better strength and to prevent disturbance of the supports.

At the time of my visit in September 1963 there were 800 feet still to be driven. **Figure 64** shows how the tunnel had to curve through 90 degrees to meet the 100 feet of tunnel previously driven from the intake chamber. Fifty feet of horizontal test boring had also been driven ahead of this short length in good rock. By September 1963 much of the face was so soft that it had to be closely boarded while 6 inches × 6 inches timber spiles had been driven over the ribs to support the roof.

The Filipinos' contract expired at the end of 1963 so there was anxiety as to whether the remaining 800 feet could be driven by that time. Much also had to be discussed over the contract settlement including the history of the Power House construction. On my return I consulted Dr Gilbert Wilson, Reader in Geology at the Imperial College. He interpreted the ground ahead from the results of two boreholes which he had asked for (A and B at 700 and 300 feet from the intake horizontal bore). His section forecast good ground for most of the way. This proved true and the Filipinos finished within their contract time.

Dr Wilson made another contribution by finding two documents by Ceylon Government Geologists in the Geological Society and Imperial College Libraries which could not easily have been available to foreign tenderers. The 1955 Report of the Government Mineralogist gave a scholarly and admirable description of the area of Stage 2 including a discussion of the 3,000 feet of diamond drilling in 17 boreholes which was not fully reproduced to tenderers.

The second was a paper in the Bulletin of the Ceylon Geographical Society Vol. 7, 1953, on *Site Investigation for Engineering Works*, which gave an excellent description of steps which should be taken, particularly with tunnels and declared the intention of putting down boreholes along the actual trace of this particular tunnel. In the event we found that borehole T.I., the only one in this vital area, instead of being bored on the trace of the tunnel was near the bottom of the valley to save the trouble and expense of climbing the mountainside. As the dip across the tunnel was 30–40 degrees down to T.I. this borehole was useless. Eventually after ten years the combined Anglo-Swedish efforts led to a successful award in an Arbitration.

Chapter 13

TUNNELLING MACHINES AND THE CHANNEL TUNNEL

The title of Tunnel Boring Machine or TBM was at first confined to full-face rotary cutters mounted within a shield or cylindrical casing. In recent years the part-face machine has come into use. This employs cutting heads mounted on booms, or other form of tool, which can wander about the face at will. It can be moved to and fro independently inside a shield or can be used with ribs and produce a tunnel with a flat floor which simplifies access to and from the face.

Between the Wars the Price Digger-Shield was the only machine successful in London Clay but the chance to use it during my time with John Mowlem & Co. did not arise. It was notable for its simplicity and ability to wander badly off line.

The period from 1958 to 1970 was spent, interwoven with many other assignments, as British counterpart to M. Rene Malcor, Ingénieur Générale des Ponts et Chaussées who was Chef de Delegation for the Channel Tunnel Study Group. The Group was made up of Directors of the British and French Channel Tunnel Companies, Heads of British and French Railways, diplomats and lawyers, French, British and American.

Our work for the first two years was to spread a small budget between a Traffic-Financial Study and a Site Investigation, which proved its feasibility, (Bruckshaw et al 1961). The details of this and the second more detailed investigation are more appropriate to the chapter on Site Investigation.

In 1882 the Lower Chalk had proved an ideal material for boring by the Beaumont rotary machine. The 7 feet pilot tunnels were driven unlined under the sea at an average rate of a metre per hour over seven months. As forecast by the Group's Consultants, the Channel Tunnel was to consist of 35 miles of twin running tunnels of 25 feet diameter, lined as driven, with a service tunnel, driven ahead as a pilot of 12 feet diameter. Since then the diameters have been much increased after further study.

The scope for tunnelling machines was considerable. For the next ten years makers were assiduous in pushing the virtues of their machines so visits were paid by me to study them in action in U.S.A., France, Britain, Germany and Switzerland. Watching others at work is no substitute for day to day working in charge of such machines but some lessons can be learned on such visits. On personal visits to

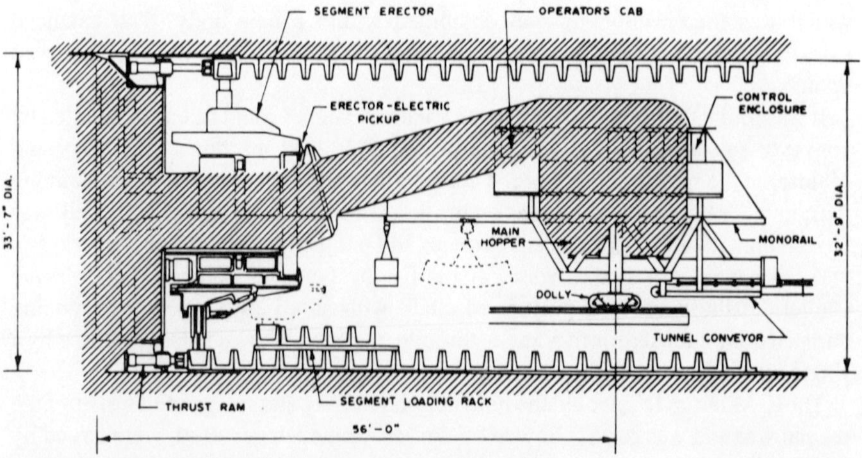

FIGURE 65. *Robbins tunnelling machine — Paris Metro extension.*

at least 14 machines only four or five were found to be at work for one reason or another.

The first, and most educational to be visited in 1959, were the 29 feet diameter Robbins *Moles* under the Oahe Dam in South Dakota, working in shales. These were the first of such large diameter to work in soft rock. My detailed report contained 4,000 words. The first machine was being dismantled, for dropping down another of the seven 42 feet diameter shafts which served seven parallel tunnels to the core of the Dam. This gave an opportunity to study it in detail.

While walking up the opposite tunnel to see the machine of the rival contractor at work, the machine came to a standstill due to a slight slip-up in the organisation. The ground was supported by circular steel ribs at four feet centres. The last visit was to the Channel Tunnel Priestley Machine in 1975 as it drove the service tunnel out to sea just before it was put into moth-balls due to the customary panic by the British Government. This beautifully made machine emphasised the accelerating sophistication of such machines compared with the effective but more basic design of the Robbins Oahe machines.

Most of the other available machines were of comparatively small diameter except for the Robbins Machine used in the Paris Metro, of 8.70 metres in diameter. **Figure 65**. The first visit was to inspect the machine, newly erected. The cutting heads were enclosed by a diaphragm so that compressed air could be applied only to the face while the pre-cast concrete segments were mechanically erected behind in free air. The spoil passed through ingenious air-lock hoppers into skips. The second visit was half-way through a successful drive but was at a standstill to overcome leakage round the tail of the shield.

The full range of data collected until 1970 is in the files of the Study Group which was then wound up and combined within a new body. The balanced assessment on tunnelling machines in *Tunnelling Technology* cannot be improved upon.

It was found that Lower Chalk in the Channel Tunnel could be easily handled by conveyor into skips but in the case of two Caldweld machines at Tilbury and Maidstone the middle chalk needed considerable addition of water to help it along the conveyor belts. At Tilbury, as on the first Dartford shield tunnel, the chalk was broken up with about 50 per cent of water and pumped to the surface.

A visit was paid to the works of the Rugby Portland Cement Works in the Chiltern Hills to observe pulverised chalk with about 40 per cent water being pumped continuously for fifty miles through a 10 inch diameter steel pipe line by a single oil-well type of pump.

The K-M Drum Digger led the field in London Clay and soft ground just after the second war and was soon followed by the McAlpine Digger. Both were issued by London Transport to several contractors for some of the drives on the Victoria Line. This led to debates on the pros and cons of leaving the choice of machine to contractors or imposing machines ordered by the client. The latter may save time but relieves the contractor of responsibilty.

Long drives under the Channel in almost ideal material would give scope for the choice of trying out two or more machines. As driving progressed it would be no great matter to discard a slower machine and replace it with a faster one. The philosophy for such long drives cannot apply in the same way to the commoner short drives in other work.

As time has gone by, the design of the *tail* towed by machines to carry all the mechanisms has become longer and longer. This can be troublesome in small diameter tunnels such as the Robbins machines visited near the summit of Mont Cenis in metamorphics and in medium limestone near Stuttgart — neither needed lining.

The use of machines in comparatively short tunnels raises financial problems. **Figure 66** shows the progress on two contracts on the same Main Drainage scheme with which I was concerned. The shield drive and the Caldweld machine worked in almost identical ground, lined with bolted concrete segments. The shield drive covered 2,040 rings at an average of 40 rings per week (including burning down the compressor shed!).

The second contractor tendered to use a shield and then asked the Engineer's permission to use a Caldweld machine, which cost eight times more than the shield. When all went well the machine averaged 70 rings per week. It took a long time to erect and a two feet thick bed of rock in the clay reduced the best average. Both tunnels were in compressed air. The longer the drive the better for the machine.

In studying the results of various machines in various works and conditions it is

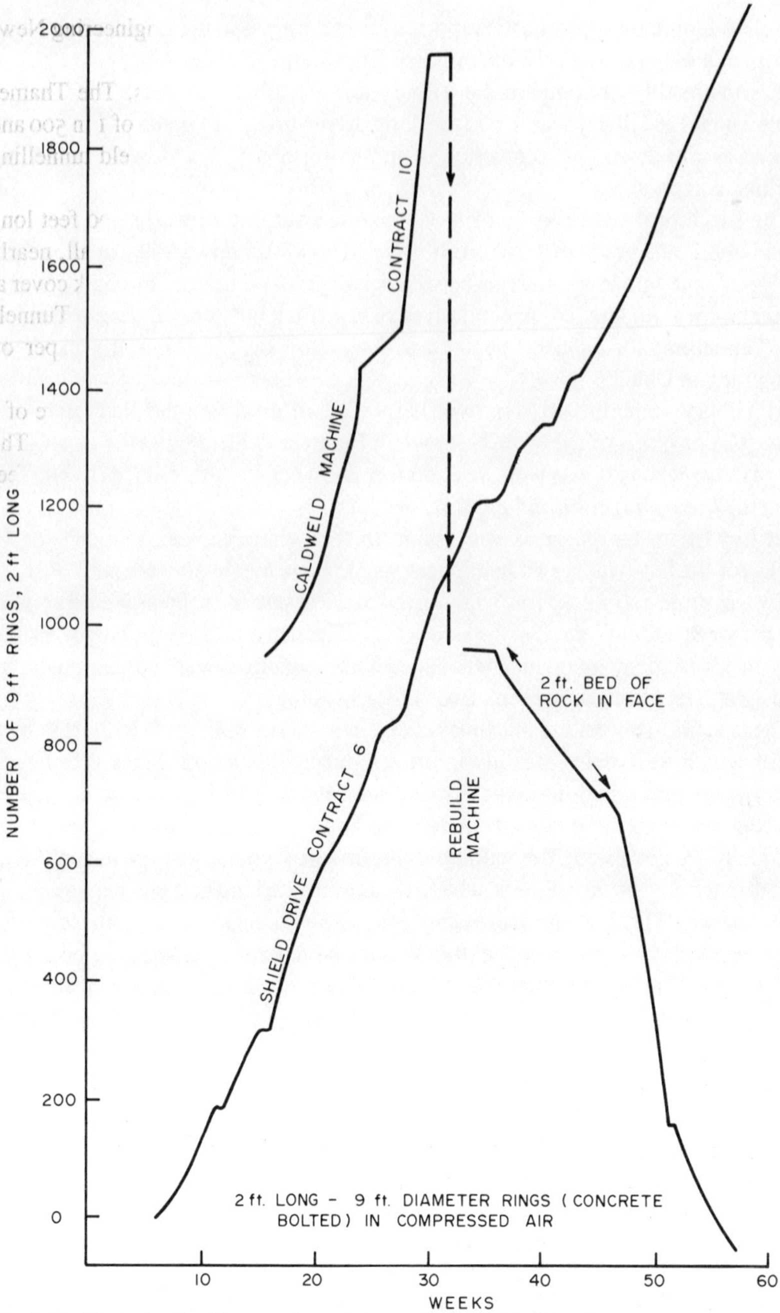

FIGURE 66. *Chart showing Comparative progress — Shield and Caldweld machine in same ground.*

as well to ignore the enthusiastic reports of best progress in the Engineering News Record and only consider the overall time from start to finish.

It is interesting to compare the Tilbury and Portobello tunnels. The Thames Cable Tunnel at Tilbury was 5,200 feet long driven uphill at a grade of 1 in 500 and took 62 weeks from the completion of the North Shaft. A Caldweld tunnelling machine was used.

The Brighton-Portobello Tunnel, as a main sewer outfall was 7,600 feet long driven uphill at a grade of 1 in 100. It was driven *by hand* and took, in all, nearly five years, but with long intervals between changes of contract. The chalk cover at the far end was 40 feet. No paper has yet been written but items appear in Tunnels and Tunnelling for January 1976 in a discussion on Mr Haswell's paper on Tunnelling in Chalk.

At Tilbury, hand drives from two shafts were offered with the alternative of a *Mole*. The progress of the Mole is shown in Figure 6 in Mr. Haswell's paper. The rate of drive when all was well was 200 feet per week but this reduces to 100 feet when the time of erection and grouting is added.

At Lewisham the progress was about 70 feet per week, so, allowing for 13 weeks for the headwalls, two hand faces would have met in 50 weeks.

If we assume 220 weeks for Brighton the average rate of progress would be 34½ feet per week and is plotted on the same chart. The machine does the hard work but may need as many men in attendance. The economics can be examined by hindsight. The events are speculative before working.

These visits to tunnelling machines ended in 1970 and only apply to the few rival makes which were being steadily improved during the period. Since then newer makes have multiplied dramatically and are recorded in articles and advertisements in magazines like *Tunnels and Tunnelling*.

Equally remarkable is the willingness to invest so much money in such individual machines of complexity which would have astonished our parsimonious predecessors. There is the increasing endeavour to make ever more powerful machines with newer types of cutter to bore harder rocks in order to eliminate drilling and blasting. All these call for hard financial decisions and in due course comparative studies of success or failure to achieve predictions would be instructive.

Accounts of record progress are encouraging but comparative studies are also needed. The genius of the mechanical and electrical engineers in conceiving and constructing these remarkable machines must be balanced by the experience of the tunnel engineer in their actual use.

The length of tunnel may be financially decisive. The longer the tunnel the more attractive is the use of the right machine. With short tunnels there is the hope of second or third uses. We need accounts of experience in transport and storage, deterioration if hope is deferred and lessons learned in rebuilding after lapse of time. Lessons learned in the previous period were the problems of conditions for erection, the inevitably slow starts, the effect of linings or their absence and the disposal of the spoil, among other factors which must be considered in assessing rates of progress.

The increase in the types and numbers of machines has been so impressive that the whole aspect of tunnelling has been changed in many respects in major works.

Chapter 14

TUNNEL LININGS, THRUST BORES AND MINI-TUNNELS

The Contractor must use the linings which are specified, so he is not usually involved in theoretical design. When looking back over 50 years, the recent acceleration in new designs and types of lining is notable. For many years there was no reason to go beyond the proven and highly adaptable cast iron lining, with lead caulking when needed for watertightness. Their long life was proved during the repairs to the Tower Subway when the original segments were taken out after 70 years and we found them to be in first class condition.

When the cost of iron and lead rocketed as the second war approached, concrete *imitation* flanged segments, with asbestos-cement caulking material proved a workable but inferior substitute. All the modern materials under the name of *plastics* had hardly come into infancy at that time. It was many years before segments made of steel plates with flanges welded on became popular in some conditions.

The real break-through and proliferation of other linings, flexible jointed, smooth surfaced and other variations started about 1956, at the same time as my own change-over from action to advising. There was a period of prosperity between slumps at that time which may have led to greater courage in investing in new developments.

There is no place for a dissertation on all the many ways of lining every sort of tunnel from rock to soft, which are as innumerable as the factors which affect decisions. For soft ground tunnels there is once more a call for standardisation, such as that of the Construction Industry Research & Information Association for a range of 10–14 sizes for circular tunnels.

That is not new. By 1922 the London Electric Railway Company had standardised 25 sizes of cast iron lining by inside diameter and valuable tables were in use for quantities per lineal yard and quantities per ring for excavation, weight, bolts, super yards of grouting and lineal yards of pointing or caulking. A third table gave every possible dimension in 16 columns.

Unfortunately the London County Council, the next largest of users for their main drainage, insisted upon their own individual sizes with different dimensions and bolt circles. Already there is a conflict between whether the inside or outside

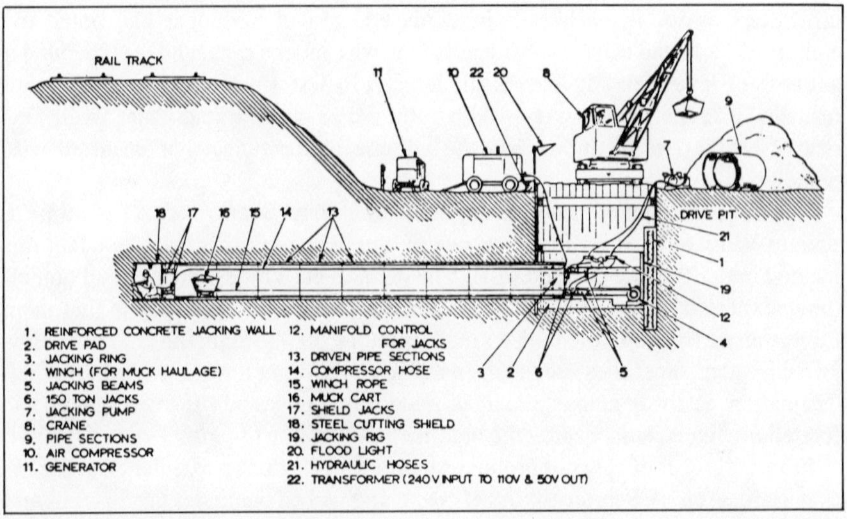

FIGURE 67. *Pipe Jacking.*

diameter should be the basis and is influenced by the use or the method of construction.

Personal observation leads to the feeling that permanence is the most important factor. Reckless economy is as bad as reckless extravagance and engineers must impress this upon the lay and inexperienced bodies who make the final budget decisions. Personal experience in altering existing services, repairing those damaged by air raids and noting collapses, emphasises the disruption which can be caused to the public who have to rely upon such services for their daily life and health.

Owing to the uncertainties below ground, a lining should be chosen which, if possible, should not need to be changed in size at a crisis, but can continue for the chosen distance. The outside diameter is thus dominant when using a shield or machine. Rock tunnels have more flexibility in these matters. For tunnels to carry liquids the modern rings giving a smooth inside surface are a good step forward compared with the expensive methods of filling inside segments with flanges.

A change in the method of lining arose with the development of pipe jacking or thrust boring in the U.K. in the 1960's, with miners working in the face. **Figure 67**. The ability to jack pipes or tubes for long distances first met with scepticism by those who were used to the forces needed to jack forward a Greathead shield. The benefit is a form of instant lining. The method was escalated to the use of large reinforced concrete boxes in appropriate places (Clarkson & Ropkins 1977).

An earlier form was the horizontal earth boring mechanical auger, working inside a jacked lining. Two rival firms, each using the method, called me in when difficulties arose. In each case in sands and gravel the auger had bored too enthusiastically and caused subsidence. One was under a canal and the other under a four track level crossing and also under a 42 inch storm culvert, a 16 inch water main and a 12 inch foul sewer — both bores were below ground-water level. This risk of boring out far more than the volume of the tunnel has occurred with tunnelling machines.

Other experiences in thrust boring include, among others, advice from time to time to William F. Rees Ltd., who were one of the the first firms to adopt this method under the title of Seerthrust. The method has virtues, but as in all tunnels the unexpected happens. From these various methods it was notable that more trouble arose from sinking the jacking pits than from the tunnelling. Those below ground-water sometimes needed deep well ground-water lowering to achieve their destination. Railway embankments provide surprises, as no site investigation can foretell the items jettisoned by the old time navvy when tipping.

From a number of cases of problems in thrust boring one particular example has a special lesson. A Seerthrust tunnel was being jacked out under the Sussex Ouse below Lewes for the South Eastern Gas Board, at Southease. The three feet diameter concrete pipes in lengths of eight feet passed eight feet below the river bed and the tidal depth varied from ten to twenty-five feet. The final length was 272 feet.

There was a jacking shaft on the west bank, and because of the pipe laying contracts the reception shaft on the east bank was the work of a rival contractor. In the jacking shaft, troubles arose towards the bottom of the steel sheet piled shaft when swelling clay or *bulls-liver* threatened trouble. On my advice the final bottoming-up was carried out with every precaution, using extra timber frames and by excavating in small sections with the centre weighted down.

This advice was passed on to the young engineer working on the east shaft in the same ground, but the latter's superiors derided such precautions. When the pipe was three-quarters across the river, in the middle of the night the miners heard and felt what they thought was an explosion. They found that the steel sheet piling in the east shaft had collapsed by the long side walls coming together so that the top of the shaft had the shape of an hour glass.

When the friction becomes too great it is possible to introduce additional thrust at a distance along the pipe. But there have been times when artificial obstructions or too much friction have forced further advances to depend on reverting to segmental tunnelling. The method has great advantages for short drives but the jacking pit itself limits length to one face, and if two faces are to meet the steering becomes less flexible than with a shield.

It is due to these varying factors that Mr Donald Rees developed his Mini-tunnel. This is a carefully worked-out method of driving the small diameter tunnels needed for sewers and water, giving longer drives and avoiding some of the pitfalls which he encountered in thrust boring. The moral is that there is no simple answer to tunnelling, but that it is a matter of infinite variety undreamed of by the layman and the administrator.

Chapter 15

SITE INVESTIGATIONS AND THE CHANNEL TUNNEL

Although site investigation is the prime necessity for a tunnel, it has been left until last as events described in the previous chapters already provide much material appropriate to the subject.

In the 1920's site investigations were performed out of sight by the semi-literate labour of well-boring firms, sometimes by percussion rigs, pounding the ground and pouring water to help to bail out the results, or by simple shell and auger in much the same way. Such samples as were seen had dried out in partitioned wooden boxes, labelled by primitive hands, apt to be upset and replaced in the many compartments. A boring log, if provided, consisted of a list of strata at depths below an unspecified ground level. It was no surprise to hear respected elderly consultants say, in discussions, that results were so misleading that it was safer not to bore at all.

In the 1930's, when called in to rescue situations by some geotechnical process, the lack of data on the ground to be dealt with was so serious that a start was made in doing one's own borings.

Those of us who served from 1943 to 1950 on the Committee which drew up the first Code of Practice for Site Investigation (CP 2001) bridged the gap between the primitive and the modern sophisticated methods.

In Britain the catalyst was the collapse of a reservoir bank at Chingford being built by John Mowlem & Co. They called in Dr Terzaghi, who supported the view of the young pioneers at the Building Research Station that it was the fault of design. This led to the establishment of the first *commercial* soils laboratory. Then followed the formation of Soil Mechanics Ltd., and the privilege of being accepted as friend and disciple of Dr Terzaghi. (Harding 1952).

When the war ended, Soil Mechanics Ltd. were the first British firm to undertake site investigations on the modern Terzaghian methods, and at the same time develop other geotechnical processes. As major engineering works were scarce at that time it was possible for the first time to have engineers who had actually run contracts supervising large site investigations and so appreciated what information really mattered. Another advance was the recruitment of young geologists who over the years produced a new breed of experienced engineering geologists. So for ten years the writer was simultaneously engaged in helping to run Soil Mechanics

Ltd., while at the same time responsible for many of the parent company's main civil engineering contracts on land and water, including tunnels.

The use of observation wells by leaving 2 inch diameter perforated pipes in boreholes before withdrawing lining tubes was learned with the ground-water lowering process and incorporated in para J. 501 of the Code of Practice. The importance of ground-water levels led to my paper to the Second International Conference on Soil Mechanics and Foundation Engineering in 1948 on "Some recorded observations of ground-water levels related to adjacent tidal movements". This included examples that the ground-water level with a tidal range of 25 feet only rose and fell a few feet in permeable ground and depended on the grain size and the position of the work. For tunnels in Thames gravels the pressure required was less than the theoretical pressure: "as the tide rises, the pressure builds up in the tunnel without special adjustment by the compressor driver".

During the post-war period our major investigations for tunnels were those for the Scottish Hydro-electric programme. Diamond drilling, which before had been the speciality of the mining engineer, was extended with warnings to clients to avoid the trap of mistaking large boulders for bedrock. Conservative consultants had to be persuaded of the value of raking boreholes to intercept fissures.

The usual reluctance of clients to pay for investigations had to be overcome. They clung to the hope of cheaper methods to avoid the cost of boring, such as probing and dubious early geophysical methods, uncalibrated by borings.

In 1945 the number of engineers with some basic knowledge of soil mechanics was small. Thirty years later almost every university or engineering college has its own soil mechanics department headed by a professor or a lecturer. Plant equipment and methods have changed rapidly. The number of firms now offering to bid for site investigations emphasises the danger of letting cheapness override experience. Some reports reveal that young men with much theory but little practice sometimes recommend the use of an expedient of which they have had no practical experience.

The previous chapters have given examples of sites for tunnels being moved away from the site investigated with no further effort to examine the new site. This is not peculiar to tunnels but also happens in dams and similar cases. Another case was a sewer tunnel to be driven along the bottom of a typical South Wales valley in the soft valley deposits with a line of borings on the lower side of the tunnel trace. The decision was then made to move the sewer laterally at the same invert level. This moved it into the mountain side and not surprisingly it became a rock tunnel and a source of dispute.

A local authority engineer asked if he could successfully sue a firm because the ground revealed in his sewer was different from that in the boreholes. On finding that the boreholes were on a parallel line, but many hundreds of yards away he was persuaded to abandon his case.

Boring over water for the line of a future tunnel dates back to the elder Brunel, whose experience of erroneous information for his Thames Tunnel is described in part 1. In rivers and sheltered waters, drilling from floating pontoons has long been the custom. The search for buried channels in river beds is now recognised as a special hazard. But handling pontoons in narrow waters with shipping lanes, rapid currents and high tidal ranges can produce difficulties.

Such a problem arose when I was asked to help in choosing alternative locations for a large main drainage sewer to cross under the famous Clifton Gorge on the River Avon at Bristol, where boring from floating craft was out of the question. A large diameter shaft was found to exist on the Bristol bank deeper than the river bed. The Bristol City Engineer was persuaded to engage the Cementation Company to bore horizontally from it with a diamond core drill along the crown of a possible tunnel and to match it with a flatly raking bore from the opposite bank, The cores proved good rock the whole way with no buried channel. The bores were grouted up as a precaution. The presence of the shaft helped to decide upon the point of crossing.

At the bottom of an oil well lies a rich reward, so money is available to build the expensive equipment evident in recent years. This includes special drilling ships and jack-up platforms. To hire these for possible tunnelling work poses severe financial decisions. This brings us inevitably to the site investigations for the Channel Tunnel.

THE CHANNEL TUNNEL SURVEYS

The make-up of the Channel Tunnel Study Group is described in Chapter Thirteen. The first needs were the financial and traffic study and the geological feasibility study. These were put in hand and organised by M. Rene Malcor as Chef de Délégation. The first site investigation study has been described in detail (Bruckshaw et al 1961). This proved that the tunnel was geologically feasible.

The hope was that measures would be taken by the British and French governments to make suitable treaties so that the necessary financial steps could be organised to raise the finance to carry out the tunnel by private enterprise. It had been the intention to carry out a much more detailed investigation as part of the main contracts during the preparations for the start of the work.

The Governments spent four years in brooding on the matter. They were told of the intention of the Group to go ahead on condition that, if once more the British Government decided to stop the work after it had got going, the Group should be compensated. The two Governments then decided to finance the main investigation themselves but for it to be done by the Study Group's engineers with a Franco-British Commission de Surveillance to see that every penny was correctly and economically spent. This much more extensive work is described by Grange & Muir-Wood in their paper to the I.C.E., January 1970 entitled *The Site Investigations for a Channel Tunnel 1964–65*.

The 1958–60 Survey

The financial resources of the study Group were limited, so hard pressure had to be exerted to extract a sufficient budget for our purposes. When we explained what we intended to do, one member of the group whose life had been spent in *administering* railways (but with a classical education) asked, "Why do we need borings? We have our experts". Our intention was to base the work on one of the marine geophysical methods just released from secrecy by the U.S. Navy, such as Sonar, Sparker or Sonoprobe. The whole area would be quartered by such methods and the results calibrated with the minimum of borings at sea as well as on land.

The tracks and locations were recorded within reasonable limits by using the existing Decca Grid system which is printed on the Admiralty charts and available to all sea-goers in the area.

The heavy Cambridge University drop-sampler was used, of the same type as that designed by Henry Brunel and used in the Channel in1866. We also had the steel engraved geological maps produced by the French engineers who had carried out their two year survey using a similar drop sampler in 1875–76 and the samples themselves: many hundreds of them were discovered in the waiting room of an abandoned French railway station. Numerous tests were carried out in the boreholes such as resistivity, self-potential and porosity by using a Schlumberger portable automatic logger. The examination of micro fossils defined a number of layers in the Middle and Lower Chalk. It was known from the exposures in the cliffs that the lower layers of the Lower Chalk were ideal for tunnelling purposes and that this overlay the Gault Clay. Fortunately the Upper Greensand is not present in this area.

Most of the offers to bore at sea, eleven in all, would only quote a price per day and offered a variety of expensive methods. As described in my paper (Harding 1949) on Site Investigation I had pioneered in the use of self-propelled tank landing craft for boring at sea at Banias off the coast of Syria. This had been noted by George Wimpey Central Laboratory who had used a schooner as a platform on the same coast and had been cast ashore. So, at this critical moment, Wimpey's were boring off Dungeness from a tank-landing craft to test the sea bed for cooling water tunnels. This was shown by me to M. Malcor and at once a simple contract was made with them to proceed to Dover.

The borings at sea, using 6 inch diamond drills on a platform cantilevered over the side produced a series of satisfactory cores. Between April and September 1959, out of a total of 140 days, 27 days were spent in boring 8 holes, plus two abortive ones, 24 days in setting the large Trinity House light buoys and moorings, while weather was too bad for work on 77 days. A sea-going launch of 13.7 metres helped to handle the moorings and carried stores and personnel to the ship. Notices to mariners of intended position had to be issued as at peak tide 100 vessels pass through the Straits in an hour. The sea bed is festooned with cables and wrecks, so the choice of position is important.

There were two geophysical surveys. In the first year 1,000 kilometres were covered by a disused lifeboat towing the Sonar equipment. The results were not as good as had been hoped, so in the next year the Sparker system was used for a month. Five days were spent in installing and dismantling plant and Decca equipment, 13½ days in surveying 500 kilometres of criss-cross routes and 11½ days lost through bad weather in April 1959.

The only tunnelling operation was the cleaning out and examination of the 1½ kilometre tunnel driven under the sea at Sangatte by the Beaumont tunnelling machine in 1882: unlined and 2.14 metres in diameter. The tunnel, above water level in the Abbotscliff between Dover and Folkstone, still exists as part of the drainage system of the Folkstone Warren landslide area.

Sir Winston Churchill was so worried about the possibility of Hitler trying to tunnel across to England that listening posts were established at Dover. Possibly Hitler was equally worried, for when the concrete seal had been taken off the old Sangatte shaft, it was found that the Germans had filled it up with masses of timber, pipes and large lumps of masonry. This made the cost of clearing the shaft considerable.

This shaft was 90 metres deep and 5.33 metres in diameter. The top and bottom sections were found to be lined with brickwork, but in the water-bearing Middle Chalk, the French mining engineers who had sunk the shaft, had used their traditional lining of rings of many short oak blocks forming a multi-polygonal face. The rings had been caulked with oakum. The oak was in fine condition, the oakum had perished. So submersible pumps had to discharge 1¾ cubic metres per minute to empty the shaft. The tunnel itself was found exactly as bored, standing unlined for 77 years under water, in the Lower Chalk, and practically dry after emptying.

The opportunity was taken to conduct as many doubters as possible down the tunnel before it had to be abandoned once more.

The Study Group delayed mobilising the firms of British and French Consulting Engineers who were to work with us in drawing up an estimate for the tunnels until the survey and the traffic studies were over.

This survey was unusual as it was personally conducted by men who would probably have considerable responsibility for the final construction work. There was complete understanding between Rene Malcor and myself and with Professor Bruckshaw and M. Goguel, the head of the French Geological survey. They were both Past Presidents of the International Geophysical Society, at the top of their profession and personal friends. For the work, simple contracts were made on one sheet of foolscap as those concerned were equally keen on the results. The overheads were minimal. Until British Rail lent us a disused railway coach converted to a drawing office, our site office was the bar of the White Cliffs Hotel. Never in the history of Site Investigation has so much been found for so little! This contrasts with the 1964–65 expedition.

The 1964–65 Investigation
This followed the same lines as the previous one but on a very much larger scale. It had little to do with actual tunnelling, but was essentially a Franco-British Combined Operation with the nautical element predominating, and carried out from Dover Castle by a site staff of forty. The results were printed in eight large volumes, with maps, charts, boring logs, chemical tests and tables of figures. The earnest student is advised to read the paper by Grange and Muir-Wood (1970), and the references which they give.

The whole story would need a book, but before proceeding to lessons learned, a summary of the work entailed may be useful.

Geophysical Profiling by *sparker* added 959 kilometres to the previous 1,500 kilometres available from 1959. *Boomer* and *Thumper* were used in appropriate places. Between 28 August and 16 October 1964, days worked were 31 and 19 were idle through bad weather.

Borehole Summary

		Vessels	Neptune Platform	G.E.M. 111 Platform	Total
For bored tunnel		44	5	20	69
For immersed tunnel		2	—	2	4
		46	5	22	73
Land bores	French 9	British 10			19
					92

Position Fixing: Decca Hi-Fix Master station at Dover Castle, slave stations at Dungeness and Bleriot Plage, Calais, giving co-ordinates within 10 metres for vessels fitted with Hi-Fix recorders. Precise survey by theodolite and Hydrodist from Ordnance Survey stations.

Immersed Tunnel Studies: on winds, waves, tides, currents, transport of sediments, movements of underwater sand dunes, echo soundings, relative sea levels on either coast, etc.

Morphology and Stratigraphy: Palaeontological zones in Lower Chalk and Gault Clay by samples at one metre intervals from 32 boreholes "using a system developed by Mr D.J. Carter of counting planktonic/bethonic ratios of foraminifera".

In-situ and Laboratory Tests: on calcium carbonate content, mechanical properties, triaxial tests, water content, density, slaking tests, etc.

Tests in Boreholes: permeability by *pumping in* through packers at various levels and pumping-out in land boreholes. Many other tests by gamma, gamma-gamma and neutron radio-metric, potential, temperature, verticality, acoustic and other tests carried out on completion of bores while ships were still on station.

The estimated cost of the investigation was to be paid for by the French and British Governments, so they appointed a Commission de Surveillance to watch over and pay for the work. It consisted of Civil Servants, both administrative and scientific. As Muir-Wood describes, this was in stark contrast to the arrangements for the 1958–60 survey. This time the Study Group were clients to their own consulting firms, but suppliants to the governments, so the administrative machine was complex.

As the work was to be far more complex and extensive than before, the five consulting firms were fully mobilised from the start to direct, organise, record and interpret the results. **Figure 68**. Malcor and I, as Delegates to the Study Group were in the position of Ring-Masters to co-ordinate the performers appearing in this Circus, even if we could not perform some of the tricks of the additional specialist experts. We two were the firm's recipients of the barracking from the audience of Commissioners at the start and in the end, shared the applause which all merited.

The Officers' Mess at Dover Castle was made available for site offices and the large mess room accommodated the 6,000 metres of cores (with 85 per cent core recovery) and also a well equipped laboratory. The office was run as a complete Franco-British unity with jobs shared out by expertise and not nationality.

The human factor refused to be ignored. Relations between individuals of the Commission and ourselves were personally harmonious but stresses arose especially in the early days. The British Chairman, the late Colonel McMullen was always courteous, but as Chief Inspecting Officer for British Railways, some of our meetings began to resemble accident enquiries. Consulting Engineers are not used to being *defendants* so I reminded them that they now knew what it was like to be a contractor.

The Commission required regular daily and monthly reports, monthly meetings in France or England, prior notification for sanction of any modification to the study and adherence to a complex accounting system. This Surveillance was so conscientious that, in the end, the Permanent Secretary to the Ministry of Transport had to appear before the Parliamentary Estimates Committee to explain why the costs had been exceeded.

The result of the Surveillance was akin to pulling up plants to see how they are growing. The Consulting Engineers were experienced in supervising works of much greater magnitude, far beyond that of most of the Commission members. The work would have been cheaper if we had been left to ourselves to make decisions and to make our own reports and accounts at suitable intervals. Elaborate charts and tables were called for every month, whereas for our own purposes one sheet of foolscap would suffice at the moment with formal charting towards the close.

The intention was to start work during the summer as the sea borings would control the time of the survey. But when at last the governments had appointed their Commission members, the latter took so long vetting every word and detail of

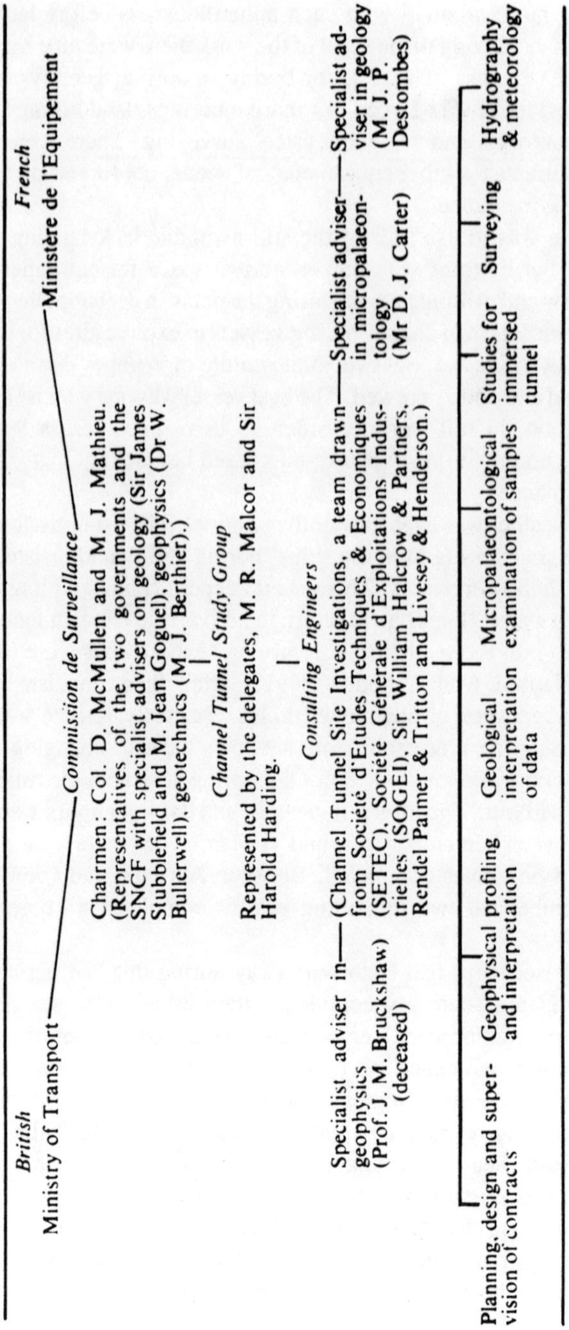

FIGURE 68. *Channel Tunnel Site Investigation, 1964–65—Organisation Chart (I.C.E. Grange & Muir-Wood.)*

every contract, large or small with such punctiliousness before letting, that the winter overtook us. Owing to the scale of the work there were nine major contracts for chartering each class of vessel, for boring, mooring, geophysical work and supporting roles, for Hi-Fix Decca, for marine borings, land borings and later for boring from platforms and for geophysical surveying. There were also twenty three minor contracts for other equipment and needs, not forgetting the important matter of marine insurance.

Our intention was to use four of the still available tank-landing craft, which provide stable flat bottoms with ample storage space for equipment and living quarters for crew and drilling gangs. During the delay in decision, the North Sea oil rush started, with a dash to charter boring vessels to explore sites for oil platforms. Two good craft were found, run by a joint venture of Wimpey-Forasol. Two other craft were found and poorly crewed. The best vessel was one which I had bought in 1950 for £10,000, in full working order, to drive piles across Spithead as an anti-submarine boom. Wimpey had re-purchased her and as G.W.14' she put up the best performance.

The winter weather was even nastier than normal. The Commissioners, without our experience, had less faith in the ships' boring ability and insisted upon work starting in the shallower waters. This meant deeper drilling and longer on station than in the deep water. Ships had to return to harbour, leaving 12 inch lining tubes bored as guides into the sea bed. On returning several tubes were found to have broken and collapsed. Muir-Wood has told in detail the mechanism of failure and the method of recovery of abandoned holes. Basically fatigue was caused by frequencies arising from the oscillations set up by the ever-changing currents.

The heaviest cost in such work is not the boring at sea but the time spent when *not* boring, in harbour. The effect of weather and daylight hours were important. **Figure 69**. Between September 1964 and 31 March 1965 there were 455 ship-days available and 8 boreholes completed. Between April 1st and completion of the work in September the two remaining vessels completed 38 boreholes in 346 ship-days.

The two inferior ships had been sent away during this bad period but, as the Commission did not share my confidence they asked us to search around for drilling platforms, rather than to engage extra ships for the summer season. Such platforms were few and expensive to hire with the added need to employ tugs to tow them from site to site. Neptune was a triangular oil drilling platform, just launched at Rouen and came for a month's shake-down for the drilling crew before going to the North Sea. The triangular lattice legs were 76 metres long with 10 metre sides. It was possible to work in 46 metres of water at low tide. Using its ponderous drilling rig, five boreholes were made.

The G.E.M.111 platform was the sawn-off end of a larger one, the Sea-Gem which collapsed in the North Sea through brittle failure of tie bars of the De Long

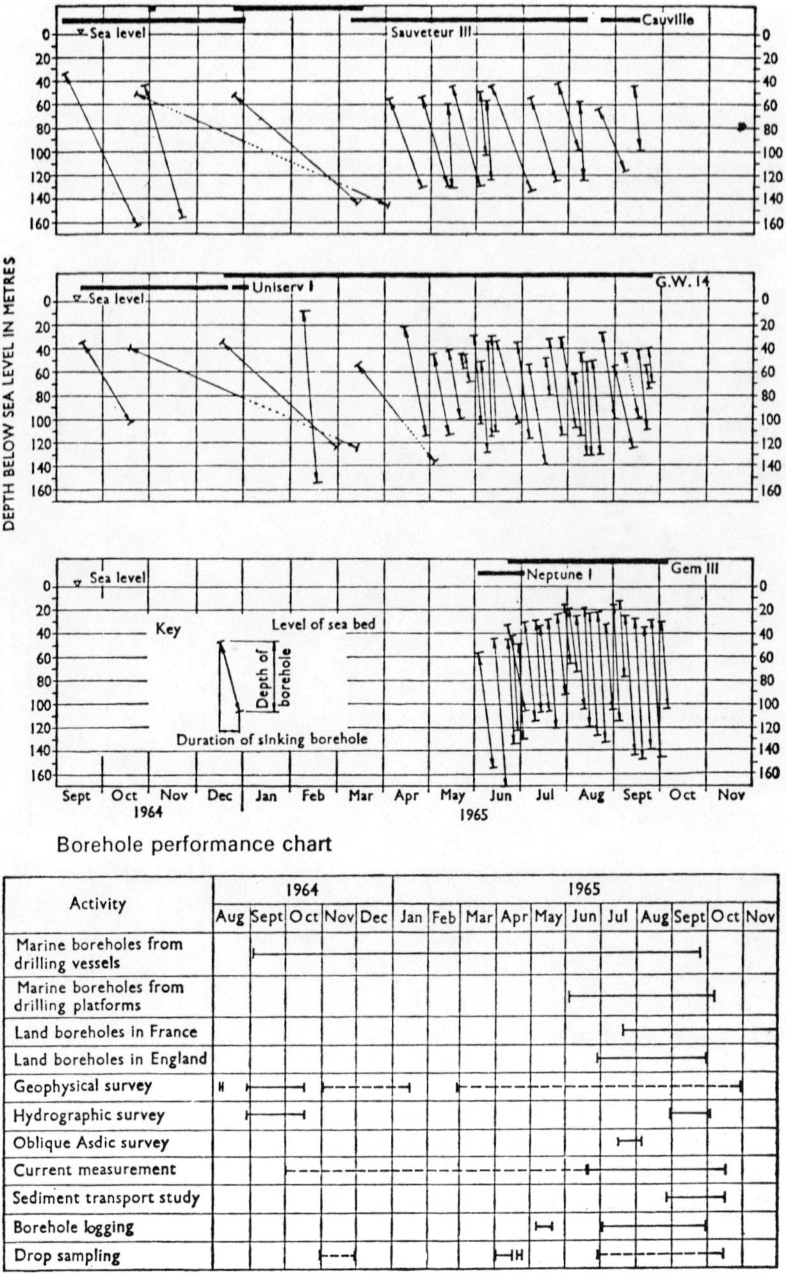

FIGURE 69. *Channel Tunnel Site Investigation — Progress Chart. (I.C.E. Grange & Muir-Wood.)*

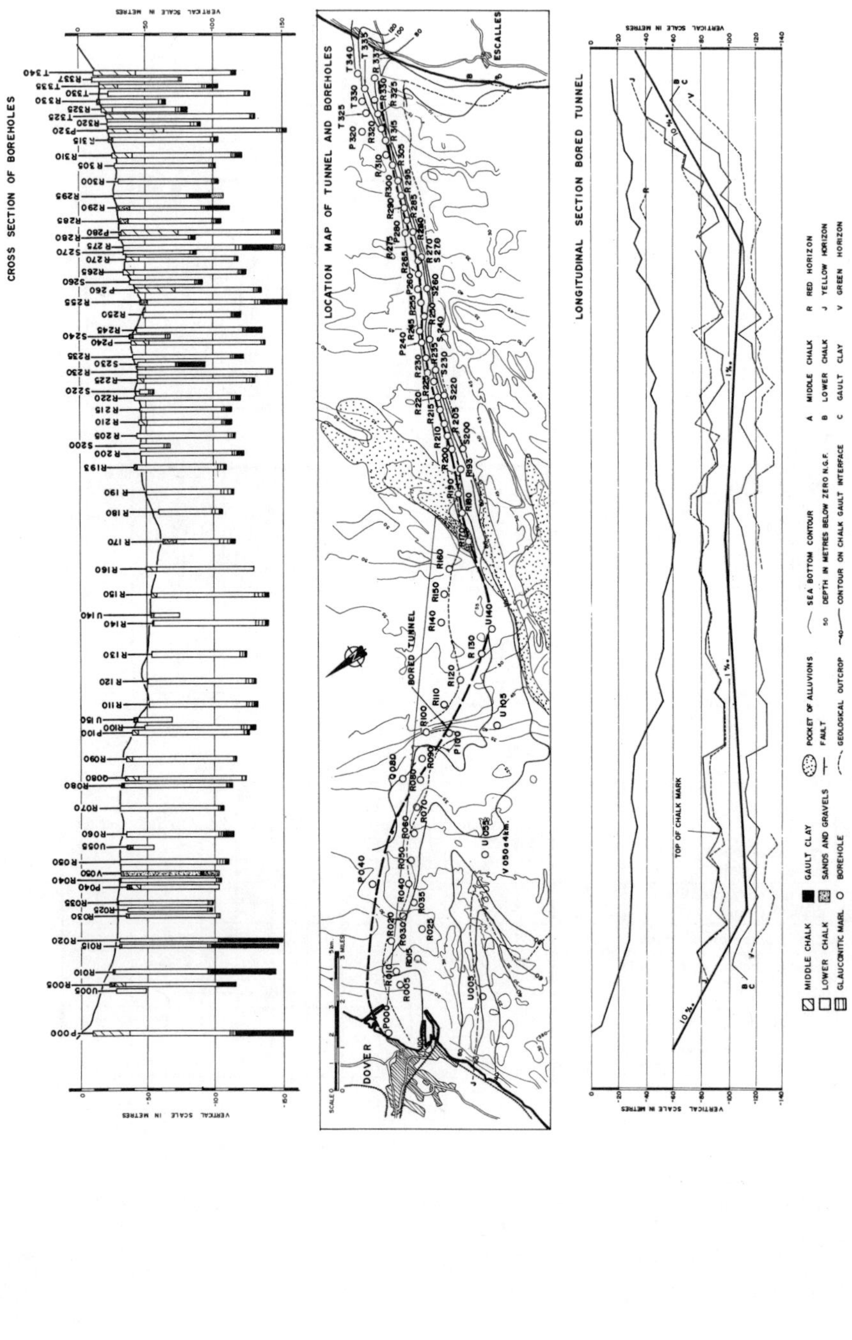

FIGURE 70. *Channel Tunnel — Cross-section of boreholes, Plan and Longitudinal section of bored tunnel.*

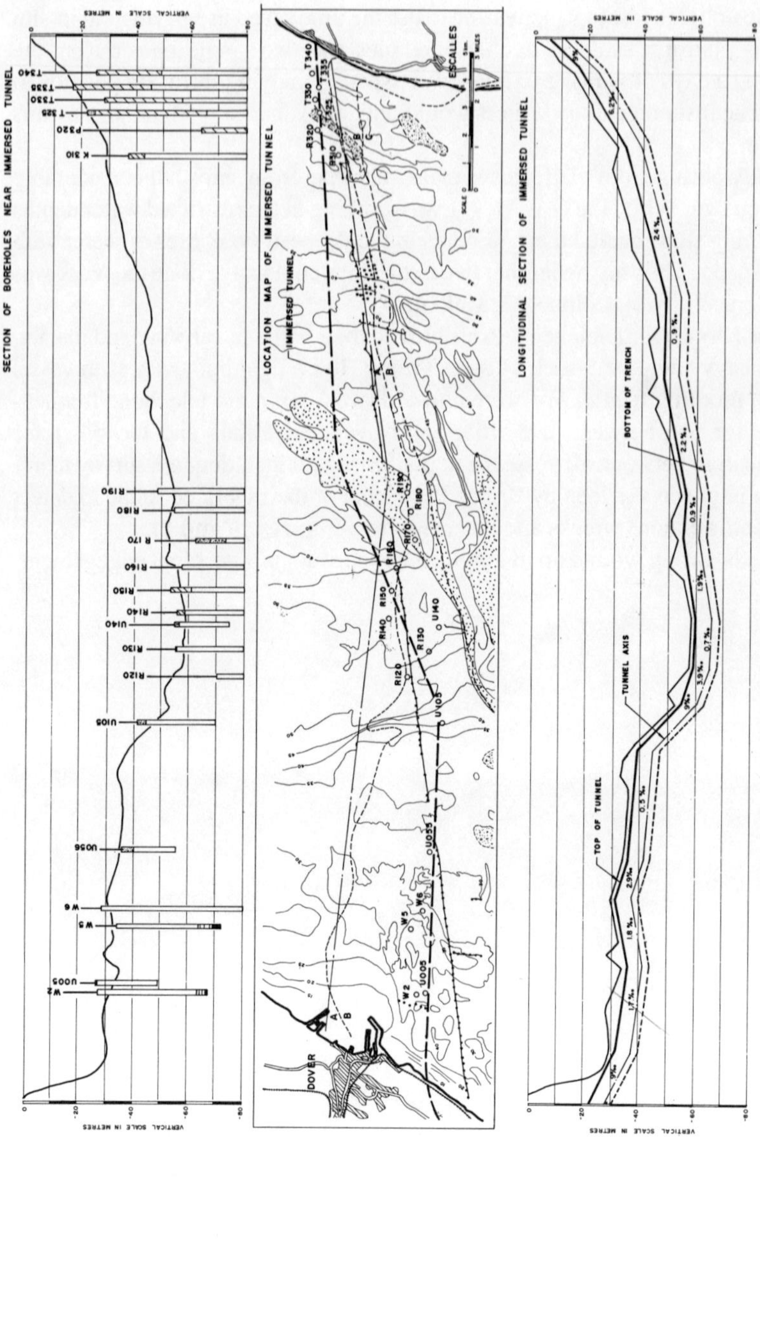

FIGURE 71. *Channel Tunnel — Boreholes, plans and sections for Immersed Tunnel.*

pneumatic jacks. Advantage was taken of its size 50 metres × 27 metres, to mount a drilling machine on each side. Two holes could be drilled at the same time. One sank a continuous cored borehole, the other used rock-roller bits to drill as quickly as possible until the last few metres and to use the time saved in carrying out in-situ tests. This platform sank 20 double boreholes and two single ones during the period of 110 days of which 230 hours were spent in moving from one position to another. The insurance underwriters would only allow her to work in 30.5 metres at low tide.

Naturally boring from platforms is pleasanter than from ships but considerably more expensive. While the G.E.M.111 was working in the restricted water depths the remaining ships drilled their 38 bores in the deepest, 60 metres of water with ease and speed. Thus my belief that two extra ships could have done the work was admitted gracefully by Colonel McMullen.

Two ex-Navy boom defence vessels proved their value in carrying and placing the very heavy equipment such as three Trinity House light buoys for each vessel and much mooring tackle. We were allowed two VHF radio telephone frequencies, one for use between the offices at Dover and Calais and the other for communication between all vessels and platforms and including the survey team. One extra item was the loan by British Petroleum of their Oblique Asdic to detect by a sweeping action, wrecks and other features. **Figures 70 and 71**.

The Study Group wound up in 1970, and so ended one's personal engagement.

ACKNOWLEDGEMENTS

Some books are the product of Inspiration. This book is the outcome of Instigation – when Doctor Golder and his Associates asked the Author to write this historical memoire with a slight philosophical bias and then to follow it up with examples of lessons learned during nearly sixty years of professional experience.

Acknowledgements are essential for gratitude, for courtesy or even for self-protection. From the study of the books of others these appear to vary from the general to the highly detailed.

Much of the experience has been gleaned many years ago with organisations and people now extinct. Apologies may be needed for some later examples to those who have unsuspectingly provided material, so let us hope for forgiveness.

A Librarian once told me that when a new book comes out visitors are apt to turn first to the Index to see if their names appear. The names of the many friends, colleagues, Senior or Junior, contacts or opposite numbers which came to mind proved too many to mention without biographical details so let us hope that they understand that they are not forgotten.

The illustrations come from many sources and are so named where possible. Thanks are due to the Institution of Civil Engineers and their Librarians past and present, and to John Bartlett and his Librarian in Messrs Mott Hay and Anderson. Also to my co-worker on the Channel Tunnel studies, Alan Muir-Wood who also is Partner of Sir William Halcrow & Partners who inherited the mantle and practice of Sir Harley H. Dalrymple Hay, under whom I cut my tunnelling teeth. Charles Haswell must not be forgotten, neither must London Transport.

Graham Whatley who produced all the art work while we were composing the Brunel Exhibition at the Institution of Civil Engineers has extracted copies of photographs which we used and whose suppliers were thanked at the time. Finally, as I have reason to know of other Authors who are writing serious, definitive works on the whole subject of tunnelling, this work had been composed in a style which, I hope, will in no way pre-empt their studies.

At a Careers Masters Conference I gave this recipe for success to aspiring engineers – "Marry the right wife, One who is not jealous of your work". For over fifty years my wife has proved this to be true by providing my work with unyielding support so this must take the form of my Dedication.

REFERENCES

AGRICOLA, G. *De Re Metallica*, 1556

ALLIN, R.V. & NACHSHEN M. Power House Foundations and Circulating Water Tunnels at the Ford Company's Works, Dagenham. I.C.E. p. 235 (1923–33)

BARTLETT, J.V., BIGGART, A.R. & TRIGGS, R.L., The Bentonite Tunnelling Machine. I.C.E. November 1963

BARTLETT, V.F & CADWELL, W.H. The Coaling Jetty, Circulating Water System and Cable Tunnels of the Battersea Power Station. I.C.E. p. 240 (1934–35)

BERRY, C.S. and A.C. DEAN The Constructional Works of the Battersea Power-Station of the London Power Co. Ltd. I.C.E. Vol 240 1937

BINNIE, G.M. N.H. GIMSON, P.F.F. LANCASTER-JONES, and C.A. GILLOTT. The Dokan Project; the flood disposal works, and the grouted cut-off curtain. I.C.E. October 1959

BLYTH, F.G.H. A geology for Engineers Edward Arnold 1953

BODY, J.B., 'The Drainage of the Valley of Mexico', I.C.E. Vol. CXLIII, 1900

BOSWELL, JAMES. 'Boswell's London Journal 1762–1763'. (Heinemann 1950).

BRUCKSHAW, J.M., GOGUEL, J., HARDING, H.J.B. and MALCOR, R., The work of the Channel Tunnel Study Group, 1958–60, I.C.E. 1962

BRUNEL, I.K., *Letters*, Bristol University Library

BRUNEL, SIR MARC, *Diaries*, I.C.E. Library

BUTTERWORTHS. Grouts and Drilling Muds in Engineering Practice. 1963.

CAMPION, F.E. Part Reconstruction of Bo-Peep Tunnel at St. Leonards-on-Sea. I.C.E. Journal: March 1951

The Civil Engineer in War, Volume 3, I.C.E. 1948

CLARKSON, T.E. & ROPKINS, J.W.T. Pipe Jacking applied to Large Structures. I.C.E. November 1977

COOLING, L.F. and H.Q. GOLDER. The Analysis of the Failure of an Earth Dam during Construction. I.C.E. J. 20. November 1942

COPPERWAITE, W.C., *Tunnel Shields and the Use of Compressed Air in Sub-Aqueous Works*, (Constable) 1906

CURRY, G. & JONES, I.J. Enlargement of the City & South London Railway Tunnels. I.C.E. April 1927

DERRY, T.K. and TREVOR, L.W., *Short History of Technology*, Oxford Clarendon Press 1960

DONOVAN, H.J. 'Expanded Tunnel Linings', *Tunnels & Tunnelling*, March/April 1974

DONOVAN, D.T., 'Henry Marc Brunel. The First Submarine Geological Survey and the Invention of the Gravity Corer', *Marine Geology*, 1966

DONOVAN, H.J. 'Modern Tunnelling Methods', *Public Works Congress* 1968

DRINKER, HENRY S., *Tunnelling, Explosive Compounds and Rock Drills*, Wiley, N.Y. 1878

ELLSON, GEORGE. Dover Train-Ferry Dock, I.C.E. J 9 December 1937

Encyclopaedia Americana American Corporation, Darling, Conn. 1978

Encyclopaedia Britannica, Volume 27 – 'Vauban', 1911

Engineering Wonders of the World, Thomas Nelson & Sons 1909

FAWCETT, B., *Railway of the Andes*, Allen & Unwin, 1906

FRANK, SIR PIERSON. Presidential Address. I.C.E. November 1945

GAMMON, K.M. & PEDGRUFT The Selection and Investigation of Potential Nuclear Power Station sites in Suffolk. I.C.E. January 1962

GAULD, G.A. and R.J. NEWPORT. The construction of two concrete-lined tunnels for the Machkund Hydro-Electric project. South India. I.C.E. P.I.3. March 1954.

GIBB, M.E. & TANSLEY, F.J. Queenhill Bridge over the River Severn. I.C.E. December 1962

GIMSON *et al*. The Dokan Project. The flood disposal Works and the Grouted Cut-off Curtain. I.C.E. October 1959

GLOSSOP, R. and SKEMPTON, A.W., 'Particle Size in Silts and Sands' *Proc. I.C.E.*, 1945

GOLDER, H.Q., 'The Use of Chemical Injection Processes in Underpinning and Tunnelling', *Die Ingenieur* 1955

GOLDER, H.Q. & SEYCHUK, J.L., Some Problems in Subway Construction 3rd. Panamerican Conference, Soil Mechanics and Foundation Engineering, Caracas, 1967

GRANGE, A. & MUIR-WOOD, A.M. The Site Investigation for a Channel Tunnel 1964–1965. I.C.E. January 1970

GROVES, G.L. Tunnel Linings and Special Reference to a new form of reinforced concrete Lining. I.C.E. March 1943

GROVES, G.L. the Ilford Tube. I.C.E. March 1946

HALL, H. The New Piccadilly Circus Station. I.C.E. Vol 228. 1928/29

HARDING, H.J.B. Chemical Consolidation of Ground. I.C.E. Manchester Dist. October 1934

HARDING, H.J.B., 'Emergency Repairs to the Tower Subway After Air Raid Damage', *Proc. I.C.E.*, 1945

HARDING, H.J.B., 'The Choice of Expedients in Civil Engineering Construction', I.C.E. Works Const. Division, 1946

HARDING, H.J.B. The Principles and Practice of Ground Water Lowering. I.C.E. Glasgow Dist. Assoc. 1946

HARDING, H.J.B. Some recorded observations of Ground-Water Levels related to adjacent Tidal Movements. 2nd. Conference on Soil Mechanics and Foundation Engineering 1948

HARDING, H.J.B., Site Investigation including Boring and other Methods of Sub-surface Exploration. I.C.E. April 1949

HARDING, H.J.B., The Practical Training of Civil Engineers. Joint Conference of Three Institutions 1951

HARDING, H.J.B., The Progress of the Science of Soil Mechanics in the Past Decade. I.C.E. James Forrest Lecture. 1952

HARDING, H.J.B., Presidential Address, Institution of Civil Engineers, 1964

HASWELL, C.K. Thames Cable Tunnel I.C.E. December 1969

HEWETT, B.H.M. and JOHANNESSEN, S., *Shield & Compressed Air Tunnelling*, McGraw Hill, N.Y. 1922

HUNTER, J.W. H.M.S. Vernon; rebuilding shore establishment; site examination and subsequent piling. I.C.E. J. 3 December 1948.

Institution of the Royal Engineers Journal, March 1976

JONES, I.H. and CURRY, G. 'Enlargement of the City & South London Railway' *Proc. I.C.E.*, April 1927

KELL, J. The Dartford Tunnel. I.C.E. February 1965

KELL, J. and G. RIDLEY. Blackwall tunnel duplication. I.C.E. p 35. October 1966

LAROUSSE, *Encyclopaedia of the Earth*, p 404

LAW, HENRY, *Memoirs of the Thames Tunnels, Part I* 1845

MEGA, T.M. & BROWN, C.D. Mersey Kingsway Tunnel. I.C.E. March 1972

MEIGH & EARLY Fourth Conference on Soil Mechanics and Foundation Engineering. Vol. I. p 68 1957

MORGAN, H.D., HASWELL, C.K. & PIRIE, E.S. The Clyde Tunnel, Design, Construction and Tunnel Services. I.C.E. February 1965

MULLER, J., *The Attack and Defence of Fortified Places* 1770 (R.E. Chatham)

PERRY, R.P. & THORNTON, G.P. Construction of the Monmouth Tunnels in Soft Rock. I.C.E. September 1970.

PEQUIGNOT, C.A. (Editor) *Tunnels and Tunnelling*, Hutchinson, 1963

PUGSLEY, SIR ALFRED. Editor The Works of Isambard Kingdom Brunel (Chapter Two – HARDING, SIR HAROLD – Tunnels) I.C.E. and University of Bristol 1976

ROLT, L.T.C., *Isambard Kingdom Brunel*, Longmans, 1957

SANDSTROM, G.E., *The History of Tunnelling*, Rochliff, 1963

SIMMS, F.W., *Practical Tunnelling*, Lockwood, London 1944

STACK, BARBARA. *The Evolutionary History of the Tunnelling Machine*, John Wiley & Sons. 1981

TERZAGHI, K. The Influence of Modern Soil Studies on the Design and Construction of Foundations. Building Research Congress. 1951

TERZAGHI, K. From Theory to Practice in Soil Mechanics. John Wiley & Sons. 1960

THOMPSON, J.C. Horizontal Earth Boring. I.C.E. April 1967

WILSON, W. STOREY & SULLY, F.W. Compressed Air Caisson Foundations I.C.E. Works Const. Div. 1949

WILSON, W. STOREY & SULLY, F.W. The Construction of the Caissons forming the Foundations to the Circulating Water Pump House for the Uskmouth Generating Station. I.C.E. December 1952

YOUNG, DESMOND, *Member for Mexico*, Cassell 1966